SpringerBriefs in Mathematics

SpringerBriefs in Mathematics showcases expositions in all areas of mathematics and applied mathematics. Manuscripts presenting new results or a single new result in a classical field, new field, or an emerging topic, applications, or bridges between new results and already published works, are encouraged. The series is intended for mathematicians and applied mathematicians.

More information about this series at http://www.springer.com/series/10030

Alexander Stoimenow

Properties of Closed 3-Braids and Braid Representations of Links

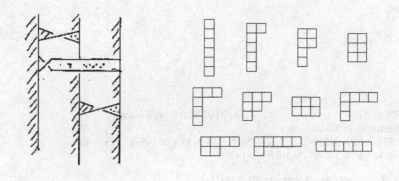

Springer

Alexander Stoimenow
School of General Studies
Gwangju Institute of Science and Technology
Gwangju, Korea (Republic of)

ISSN 2191-8198   ISSN 2191-8201 (electronic)
SpringerBriefs in Mathematics
ISBN 978-3-319-68148-1   ISBN 978-3-319-68149-8 (eBook)
https://doi.org/10.1007/978-3-319-68149-8

Library of Congress Control Number: 2017954371

Mathematics Subject Classification: 20F36, 20C08, 32S55, 12D10

Printed on acid-free paper

This Springer imprint is published by Springer Nature
The registered company is Springer International Publishing AG
The registered company address is: Gewerbestrasse 11, 6330 Cham, Switzerland

Preface

This book consists of a long manuscript devoted to the study of diverse aspects of braid representations of knots and links. We focus predominantly on 3, and partially on 4 strands, but do try to generalize some results to higher braid groups.

Here is a brief summary of the content. We show that 3-braid links with given (non-zero) Alexander or Jones polynomial are finitely many and can be effectively determined. We classify among closed 3-braids strongly quasi-positive and fibered ones and show that 3-braid links have a unique incompressible Seifert surface. We also classify the positive braid words with Morton-Franks-Williams bound 3 and show that closed positive braids of braid index 3 are closed positive 3-braids. For closed braids on more strings, we study the alternating links occurring. In particular we classify those of braid index 4 and show that their Morton-Franks-Williams inequality is exact. We settle the strong form of Jones' exponent sum conjecture for (arbitrary) braid representations of 3-braid links. Finally, we use the Burau representation to obtain new braid index criteria, including an efficient 4-braid test.

The manuscript was mainly written between 2004 and 2006, with the latest important additions made around 2009. In my previous book [St5], I attempted to extend a similarly minded research manuscript by a broadly accessible and at times entertaining introductory part and finished with an outlook exposition (citing further-going work, mostly of myself). Here I decided, perhaps somewhat extremely, to preserve as much as possible the style of a research monograph and the original state (and level and scope) of the material. Thus editing was limited to formalities like relayout, update of references, and addition of an index. Apart from correcting obvious typos I found, I have also retained the language style, including a few instances of perhaps perceivably peculiar writing, for example, the "ß" of Gauß. (While this letter has meanwhile become obsolete even in German, it is generally agreed that the so-called orthography reform has, as a whole, made German writing only more peculiar...)

I had several reasons to refrain from substantial alterations. One was that there were no changes recommended by the referee. Moreover, the publisher requested a space limitation (in the spirit of calling the volumes in the series to be "brief").

In particular, I saw no problem that all subsequent related study be also left entirely (and tacitly) to its own separate place.

In result, again (and even more so than with my previous book), the work primarily concentrates on providing complete proofs for a certain number of theorems and is not aimed to serve well as a textbook on knot/braid theory or as an overview of the topic. I realize that this makes the exposition somewhat less self-contained than some readers would desire, but I hope there are enough references given for the needed basics.

There is one noteworthy update that must be mentioned. D. Emmes in "An expression for the Homflypt polynomial and some applications," Topology and its Applications 160 (2013) 2069–2087, has fully answered affirmatively the question posed and partially settled in Section 4.2: does the Jones polynomial of a 3-braid link determine its skein polynomial? Apart for this important point (and drawing attention to the reference to Y. Ni given in the introduction), I am not (and have not been made) aware of errors found, proofs simplified, results improved, or questions answered.

Several people must be thanked for realizing this project. Crucially, I owe gratitude to M. Hirasawa, M. Ishiwata, and K. Murasugi for providing some missing pieces of the proofs, given in the appendix. I also like to thank the referee, R. Amzad, and the technical team at Springer for their support.

Gwangju, Korea (Republic of)   Alexander Stoimenow
August 2016

Acknowledgements

Apart from crediting the contribution of M. Hirasawa, M. Ishiwata, and K. Mura-sugi, I am grateful to S. Kamada, K. Kawamuro, T. Nakamura, and D. Silver for some helpful remarks. The calculations were performed largely using the programs of [MS2] and [HT], and MATHEMATICA™ [Wo]. I also wish to thank to my JSPS host Prof. T. Kohno at University of Tokyo for his support.

Contents

Chapter 1
Introduction

Originating from the pioneering work of Alexander [Al2] and Artin [Ar, Ar2], braid theory has become intrinsically interwoven with knot theory, and over the years, braid representations of different types have been studied, many of them with motivation coming from fields outside of knot theory, for example dynamical systems [Wi] or four-dimensional QFTs [Kr].

Braid representations of links are the main topic of this work. On the one hand, we will be particularly interested in the study of the link polynomials [BLM, Al, J2, F&] on braids. On the other hand, we will give closer attention to closures of 3-braids. They will be studied not only with regard to polynomials, but also to some geometric properties. For several such properties we will obtain complete classification results.

The methods we will apply are of various nature, but two stand out in importance. The earlier part of the monograph is heavily influenced by Xu's normal 3-braid form [Xu], and may be considered as a development of some earlier ideas in [St2]. The later part contains a substantial treatment of the Hecke algebra representation theory of link polynomials developed by Jones [J].

We introduce now our results in some historical context.

The systematic study of closed 3-braids was begun by Murasugi [Mu2]. Later 3-braid links have been classified in [BM], in the sense that 3-braid representations of the same link are exactly described. This is, up to a few exceptions, mainly just conjugacy. The conjugacy problem of 3-braids has a series of solutions, starting with Schreier's algorithm [Sc], going over Garside [Gr] (for arbitrary braid groups), Xu [Xu] (and later more generally Birman–Ko–Lee [BKL]), until, for example, a recent algorithm of polynomial complexity due to Fiedler and Kurlin [FK]. So Birman–Menasco's work allows to decide if two 3-braids have the same closure link.

However, many properties of links (except achirality and invertibility) are not evident from (3-strand) braid representations, and thus to classify 3-braid links with

© The Author(s) 2017

A. Stoimenow, *Properties of Closed 3-Braids and Braid Representations of Links*,
SpringerBriefs in Mathematics, https://doi.org/10.1007/978-3-319-68149-8_1

Fig. 1.1 A braided Seifert
surface obtained from a band
representation

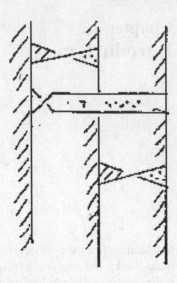

special properties remains a non-trivial task. A first result was given in [Mu] for
rational links, and then improved in [St2], where we completed this project for
alternating links.

From the opposite point of view, a natural question concerning braid representations of links is:

Question 1.1 If a braid representation of a particular type exists, does also one exist
with the minimal number of strands (among all braid representations of the link)?

The minimal number of strands for a braid representation of a link L is called the
braid index of the link, and will be denoted by $b(L)$.

In [Be], Bennequin studied in relation to contact structures braid representations
by bands, independently considered by Rudolph [Ru] in the context of algebraic
curves, and more recently in [BKL] from a group theoretic point of view. A band
representation naturally spans a Seifert surface of the link. See Figure 1.1.

Bennequin proved (see Theorem 2.2) that a minimal genus surface of a 3-braid
links can be represented in this band form on a 3-braid, and it was called a
Bennequin surface by Birman–Menasco [BM2]. Later, by examples found by Ko
and Lee, and myself jointly with Hirasawa, the existence of Bennequin surfaces on
the minimal number of strands was found not to extend to 4-braids. (We will see in
this monograph, though, among many other things, that it does extend for alternating
links, as outlined below.) Since by Rudolph's work [Ru], a Bennequin surface exists
for any link, on a possibly large number of strands, the answer to the Question 1.1
is negative for minimal genus band representations.

A special case of band representations are the positive ones, called strongly
quasi-positive in [Ru]. For such representations the above examples do not apply.
The existence of minimal such representations is a question of Rudolph, whose
answer is not known. Our first result is a positive answer to Rudolph's question for
3-braids, and is an application of the work in [St2] on Xu's normal form [Xu].

Theorem 1.1 *If L is a strongly quasi-positive link, and has braid index 3, then L has a strongly quasi-positive 3-braid representation.*

The following work involves a further study of Xu's normal form. A natural relation of this form to the skein polynomial was exhibited in [St2]. Here we extend this study to the Alexander polynomial Δ. As a result, we can describe 3-braid links with given Δ, and in particular show no non-trivial 3-braid knot has trivial polynomial. In [B], Birman constructed some pairs of different 3-braid links with the same polynomials and proposed the problem to understand the values of the various link polynomials on 3-braid links. Our work solves (to the most satisfactory extent) Birman's problem for the Alexander polynomial. We also have a solution for the Jones and Brandt–Lickorish–Millett–Ho Q polynomial.

Parallel to the Alexander polynomial description, and using a recent result of Hirasawa-Murasugi, we classify among 3-braid links the fibered ones. (This result was proved independently by Yi Ni in [Ni], apparently unaware of this work, which I did to make widely public immediately.) We will see then that a 3-braid knot is fibered if (and only if) its Alexander polynomial is monic, but that this is *not* true for 2- or 3- component links.

The fact that most 3-braid links were found fibered, and this could be proved independently from Bennequin and Birman–Menasco, opened the hope for a more natural approach to some of their results. Finally we were indeed able to complete this project, with the assistance of M. Hirasawa and M. Ishiwata, obtaining

Theorem 1.2 *Any 3-braid link has a unique (non-closed) incompressible surface.*

This result subsumes all previous uniqueness theorems. Besides, its proof is entirely different, in that it fully avoids the contact geometry approach of Bennequin and the considerable complexity in Birman–Menasco's subsequent braid foliation work. Instead, we use the sutured manifold theory of Gabai [Ga, Ga2] (to deal with the fibered links) and Kobayashi–Kakimizu [Ko, Kk] (for the remaining cases), as well as some of our preceding work on the Alexander polynomial (to rule out disconnected Seifert surfaces). Since Kobayashi–Kakimizu use a slightly stronger notion of uniqueness (see the beginning of Section 3.2), Theorem 1.2 is an improvement even for minimal genus surfaces. Moreover, its proof underscores the geometric meaning of Xu's form, that remained unclear in [BM].

The study and applications of Xu's form occupy the first main part of the monograph. Then some related results are included, whose treatment requires different methods. These results are dealt with in the later chapters of the monograph.

As a first such topic, more substantial effort will be made to prove an analogous result to Theorem 1.1 for positive (in the usual Artin generators) braid representations.

Theorem 1.3 *If L is a positive braid link, and has braid index 3, then L is a positive 3-braid link.*

So our work here can be understood to answer Question 1.1 for braid index 3 and strongly quasi-positive or positive braid links. (The status of quasi-positive links, and of quasi-positive braid representations with regard to Question 1.1, in contrast remains open. There is, however, an algorithm, found by Orevkov [Or] to decide if a 3-braid is quasi-positive.)

The proof of Theorem 1.3 requires a study of positive braids where the braid index bound in the Morton-Franks-Williams inequality [Mo, FW] is 3. This study builds on and extends (but also considerably simplifies) the work of Nakamura [Na] for Morton–Franks–Williams bound 2. Then some detailed case distinction and calculation are necessary. In [St3] it was shown, again, that Theorem 1.3 is not true for 4-braids: there are two 16 crossing knots that have positive 16 crossing 5-braid representations, but braid index 4 (and consequently only non-positive 4-braid representations). It is in fact for these examples that I was led to investigate about the theorem.

In Section 5.2, using the arguments in the proof of Theorem 1.3 and a criterion of Yokota [Yo], we consider 3-braid representations of links that are (diagrammatically) positive. We obtain strong restrictions on such representations, and in particular determine which of the links are not fibered. (See Theorem 5.5.)

Another problem proposed by Birman, in [Mo2], was to understand the relation of alternation of links (as a pre-eminent diagram-defined property) and braid representations. Early substantial results on the braid index of alternating links were due to Murasugi [Mu], who determined the braid index for rational and fibered alternating links. In [St2] we used Xu's form to classify the braid index 3 alternating links. Here, in Chapter 6, we will easily recover this result and push it forward to braid index 4. For this we use an argument based on the Jones polynomial, and connect the celebrated Kauffman–Murasugi–Thistlethwaite work [Ka2, Mu3, Th2] to braid representations. The existence of the Bennequin surface of alternating links on a 4-string braid (Corollary 6.4) is among several easy consequences we obtain.

Chapter 7 discusses some applications of the Hecke algebra representation theory of link polynomials [J]. One point of focus will be the proof of a conjecture of Jones for closed 3-braids in Section 7.1. Then we obtain estimates on the norm of special values of the Alexander and Jones polynomial coming from the Squier unitarization [Sq] of the Burau representation, that concern also braids on more strings. They relate to and partly extend estimates of Jones in [J]. In particular, we will see that the Alexander polynomial provides conditions for every given braid index. Some properties of the skein polynomial, and a relation to Mahler measure are also discussed.

In the case of a 4-braid, one can go further and (almost) identify the Eigenvalues of the Burau matrix from the Alexander and Jones polynomial (of its closure). Using this, a criterion for braid index 4 is derived. We show examples exhibiting the efficiency of this test, including such where not only the Morton-Franks-Williams inequality itself, but also its 2-cable version fails (and so our test seems the only practicable option).

In the appendix we collect some work of Hirasawa, Ishiwata, and Murasugi, which completes the proof of several of our results.

Chapter 2
Preliminaries, Basic Definitions, and Conventions

Basic concepts that appear throughout the monograph are summarized here. "W.l.o. g." will abbreviate "without loss of generality"; "r.h.s." will stand for "right hand-side". The notations ":=" and "=:" indicate definition, with the defined symbol standing on the hand-side of the colon.

2.1 Links and Link Diagrams

Links are represented by diagrams; we assume diagrams are oriented (though sometimes orientation is not relevant).

A crossing p in a link diagram D is called *reducible* (or nugatory) if it looks like onthe left of Equation (2.1). D is called reducible if it has a reducible crossing, else it is called *reduced*. The reducing of the reducible crossing p is the move depicted on Equation (2.1). Each diagram D can be (made) reduced by a finite number of these moves.

$$\tag{2.1}$$

We assume in the following all diagrams reduced, unless otherwise stated.

The diagram on the right of Figure 2.1 is called *connected sum A#B* of the diagrams A and B. If a diagram D can be represented as the connected sum of diagrams A and B, such that both A and B have at least one crossing, then D is called *disconnected* (or composite), else it is called *connected* (or prime). K is *prime* if whenever $D = A\#B$ is a composite diagram of K, one of A and B represents an unknotted arc (but not both; the unknot is not considered to be prime per convention).

© The Author(s) 2017
A. Stoimenow, *Properties of Closed 3-Braids and Braid Representations of Links*,
SpringerBriefs in Mathematics, https://doi.org/10.1007/978-3-319-68149-8_2

Fig. 2.1 Connected sum of diagrams

By $c(D)$ we denote the *number of crossings* of D, $n(D)$ the number of components of D (or K, 1 if K is a knot), and $s(D)$ the *number of Seifert circles* of D. The *crossing number $c(K)$* of aknot or link K is the minimal crossing number of all diagrams D of K.

The *Seifert graph $\Gamma(D)$* of a diagram D is defined to be the graph whose vertices are the Seifert circles of D and whose edges are the crossings. The *reduced Seifert graph $\Gamma'(D)$* is defined by removing multiple copies of an edge between two vertices in $\Gamma(D)$, so that a simple edge remains. (See [MP, MP2] for example.)

The *(Seifert) genus $g(K)$* resp. *Euler characteristic $\chi(K)$* of a knot or link K is said to be the minimal genus resp. maximal Euler characteristic of Seifert surface of K. For a diagram D of K, $g(D)$ is defined to be the genus of the Seifert surface obtained by Seifert's algorithm on D, and $\chi(D)$ its Euler characteristic. We have $\chi(D) = s(D) - c(D)$ and $2g(D) = 2 - n(D) - \chi(D)$.

We denote by $!D$ the *mirror image* of D, and $!K$ denotes the mirror image of K. Clearly $g(!D) = g(D)$ and $g(!K) = g(K)$.

Similarly we use χ_s, g_s for (smooth) slice genus and Euler characteristic.

The numbering of knots we use is as in the tables of [Ro, Appendix] for prime knots of crossing number ≤ 10, and as in [HT] for those of crossing number 11–16. KnotScape's numbering is reorganized so that for given crossing number non-alternating knots are appended after alternating ones, instead of using "a" and "n" superscripts.

2.2 Polynomial Link Invariants

Let $X \in \mathbb{Z}[t, t^{-1}]$. The *minimal* or *maximal degree* $\min \deg X$ or $\max \deg X$ is the minimal resp. maximal exponent of t with non-zero coefficient in X. Let $\mathrm{span}_t X = \max \deg_t X - \min \deg_t X$. The coefficient in degree d of t in X is denoted $[X]_{t^d}$ or $[X]_d$. The *leading coefficient* $\max \mathrm{cf}\, X$ of X is its coefficient in degree $\max \deg X$. If $X \in \mathbb{Z}[x_1^{\pm 1}, x_2^{\pm 1}]$, then $\max \deg_{x_1} X$ denotes the maximal degree in x_1. Minimal degree and coefficients are defined similarly, and of course $[X]_{x_1^k}$ is regarded as a polynomial in $x_2^{\pm 1}$.

Let $P(v, z)$ be the *skein polynomial* [F&, LM]. It is a Laurent polynomial in two variables of oriented knots and links. We use here the convention of [Mo], i.e. with the polynomial taking the value 1 on the unknot, having the variables v and z and satisfying the skein relation

$$v^{-1} P\left(\diagup\!\!\!\!\diagdown\right) - v P\left(\diagup\!\!\!\!\diagdown\right) = z P\left(\,\big)\,\big(\,\right).$$

$$(2.2)$$

We will denote in each triple as in (2.2) the diagrams (from left to right) by D_+, D_- and D_0. For a diagram D of a link L, we will use all of the notations $P(D) = P_D = P_D(v, z) = P(L)$, etc. for its skein polynomial, with the self-suggestive meaning of indices and arguments. So we can rewrite (2.2) as

$$v^{-1}P_+(v, z) - vP_-(v, z) = zP_0(v, z). \tag{2.3}$$

The *writhe*, or *(skein) sign*, is a number ± 1, assigned to any crossing in an oriented link diagram. A crossing as on the left in (2.2) has writhe 1 and is called *positive*. A crossing as in the middle of (2.2) has writhe -1 and is called *negative*. The writhe of a link diagram is the sum of writhes of all its crossings.

Let $c_\pm(D)$ be the number of positive, respectively negative crossings of a diagram D, so that $c(D) = c_+(D) + c_-(D)$ and $w(D) = c_+(D) - c_-(D)$.

The *Jones polynomial* [J2] V and (one variable) *Alexander polynomial* [Al] Δ are obtained from P by variable substitutions

$$\Delta(t) = P(1, t^{1/2} - t^{-1/2}), \tag{2.4}$$

and

$$V(t) = P(t, t^{1/2} - t^{-1/2}). \tag{2.5}$$

Hence these polynomials also satisfy corresponding skein relations. (In algebraic topology, the Alexander polynomial is usually defined only up to units in $\mathbb{Z}[t, t^{-1}]$; the present normalization is so that $\Delta(t) = \Delta(1/t)$ and $\Delta(1) = 1$.)

In very contrast to its relatives, the *range* of the Alexander polynomial (i.e., set of values it takes) is known. Let us call a polynomial $\Delta \in \mathbb{Z}[t^{1/2}, t^{-1/2}]$ *admissible* if it satisfies for some natural number $n \geq 1$ the three properties

1. $t^{(n-1)/2}\Delta \in \mathbb{Z}[t^{\pm 1}]$,
2. $\Delta(t) = (-1)^{n-1}\Delta(1/t)$ and
3. $(t^{1/2} - t^{-1/2})^{n-1} \mid \Delta$ for $n > 1$, or $\Delta(1) = 1$ for $n = 1$.

It is well-known that these are exactly the polynomials that occur as (1-variable) Alexander polynomials of some n-component link.

The *Kauffman polynomial* [Ka] F is usually defined via a regular isotopy invariant $\Lambda(a, z)$ of unoriented links. We use here a slightly different convention for the variables in F, differing from [Ka, Th2] by the interchange of a and a^{-1}. Thus in particular we have for a link diagram D the relation $F(D)(a, z) = a^{w(D)}\Lambda(D)(a, z)$, where $\Lambda(D)$ is the writhe-unnormalized version of the polynomial, given in our convention by the properties

$$\Lambda\left(\times\right) + \Lambda\left(\times\right) = z\left(\Lambda\left(\asymp\right) + \Lambda\left(\,\right)\left(\,\right)\right), \tag{2.6}$$

$$\Lambda\left(\text{⟨⟩}\right) = a^{-1} \Lambda\left(\text{| |}\right); \quad \Lambda\left(\text{⟨⟩}\right) = a \,\Lambda\left(\text{| |}\right), \tag{2.7}$$

$$\Lambda\left(\bigcirc\right) = 1. \tag{2.8}$$

The *Brandt-Lickorish-Millett-Ho polynomial* [BLM] Q is given by $Q(z) = F(1, z)$.

2.3 Semiadequacy

An alternative description of V is given by the Kauffman bracket in [Ka2]. We do not need this description directly, but for self-containedness it is useful to recall the related concept of semiadequacy that was popularized in [LT].

Let D be an unoriented link diagram. A *state* is a choice of splittings of type A or B for any single crossing (see Figure 2.2). Let the *A-state* of D be the state where all crossings are A-spliced; similarly define the B-state.

We call a diagram *A-(semi)adequate* if in the A-state no crossing trace (one of the dotted lines in Figure 2.2) connects a loop with itself. Similarly we define B-(semi)adequate. A diagram is *semiadequate* if it is A- or B-semiadequate, and *adequate* if it is simultaneously A- and B-semiadequate. A link is adequate/semiadequate if it has an adequate/semiadequate diagram. (Here A-adequate and B-adequate is what is called $+$adequate resp. $-$adequate in [Th].)

We also need a part of the semiadequacy formulas for the Jones polynomial. As in [St4], for a non-negative integer i, we write V_i for $[V]_{\min \deg V + i}$ and $\bar{V}_i = [V]_{\max \deg V - i}$. (These are the $i + 1$st and $i + 1$st last coefficient of V in its coefficient list.) For an A- (resp. B-) semiadequate diagram, in [St4], and independently in [DL], formulas were obtained for $V_{1,2}$ (resp. $\bar{V}_{1,2}$).

From the A-state $A(D)$ of a diagram D we define two graphs. The first graph, we call it the *A-graph* $G(A) = G(A(D))$ has vertices for each loop in $A(D)$, and an edge between each pair of loops that are connected by a trace of at least one crossing in D. If there are at least two such traces, we call the edge *multiple*.

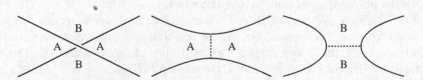

Fig. 2.2 The A- and B-corners of a crossing, and its both splittings. The corner A (resp. B) is the one passed by the overcrossing strand when rotated counterclockwise (resp. clockwise) towards the undercrossing strand. A type A (resp. B) splitting is obtained by connecting the A (resp. B) corners of the crossing. It is useful to put a "trace" of each splitted crossing as an arc connecting the loops at the splitted spot

Let $\Delta(D) = \Delta(A(D))$ be the number of cycles of length 3 (*triangles*) in $G(A(D))$. (Multiple edges do not form different triangles between the same three vertices.)

The second graph, we call it the *intertwining graph* $IG(A) = IG(A(D))$, has vertices for each multiple edge in $A(D)$, and an edge in $IG(A)$ is drawn between each pair of vertices in $IG(A)$, whose corresponding multiple traces in $A(D)$ contain traces of the form shown below.

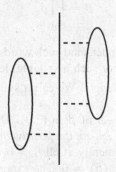

Theorem 2.1 ([St4, DL]) *Assume that D is a connected A-adequate diagram. Then* $V_0 V_1 = \chi(G(A)) - 1$ *and*

$$V_0 V_2 = \binom{2 - \chi(G(A))}{2} + \chi(IG(A)) - \Delta(A(D)).$$

Here χ is the Euler characteristic (number of vertices minus number of edges); $V_0 = \pm 1$ by [LT]. Analogous formulas hold for *B-adequate diagrams*.

2.4 Braids and Braid Words

The *n*-string *braid group* B_n is considered generated by the Artin *standard generators* σ_i for $i = 1, \ldots, n - 1$. These are subject to relations of the type $[\sigma_i, \sigma_j] = 1$ for $|i - j| > 1$, which we call *commutativity relations* (the bracket denotes the commutator) and $\sigma_{i+1}\sigma_i\sigma_{i+1} = \sigma_i\sigma_{i+1}\sigma_i$, which we call *Yang–Baxter* (or shortly YB) relations.

We will make one noteworthy modification of this notation. In the following σ_3 will stand for the usual Artin generator for braids of 4 or more strands (such braids are considered explicitly only in Chapters 5.1 and 7), while on 3-braids (considered in all other chapters) it will denote the "band" generator $\sigma_2\sigma_1\sigma_2^{-1} = \sigma_1^{-1}\sigma_2\sigma_1$.

It will be often convenient in braid words to write $\pm i$, in brackets, for $\sigma_i^{\pm 1}$. For example, $[21(33 - 4)^2 32]$ is an alternative writing for $\sigma_2\sigma_1(\sigma_3^2\sigma_4^{-1})^2\sigma_3\sigma_2$.

The following definition summarizes basic terminology of braid words used throughout the monograph.

Definition 2.1 Choose a word

$$\beta = \prod_{i=1}^{n} \sigma_{k_i}^{l_i} \tag{2.9}$$

with $k_i \neq k_{i+1}$ and $l_i \neq 0$. We understand such words *in cyclic order*.

Call $\sigma_{k_i}^{l_i}$ the *syllables* of β. For such a syllable, let k_i be called the *index* of the syllable, and l_i its *exponent* or *length*. We call a syllable $\sigma_{k_i}^{l_i}$ *non-trivial* if $|l_i| > 1$ and *trivial* if $|l_i| = 1$.

We say n is the *syllable length* of β in (2.9). The *(word) length* $c(\beta)$ of β is $\sum_{i=1}^{n} |l_i|$, and $c_{\pm}(\beta) = \sum_{\pm l_i > 0} \pm l_i$ are the *positive/negative length* of β. A word is *positive* if $c_-(\beta) = 0$, or equivalently, if all $l_i > 0$. Similarly it is *almost positive* if $c_-(\beta) = 1$ and *negative* if $c_+(\beta) = 0$. The *exponent sum* of β is defined to be $[\beta] = \sum_{i=1}^{n} l_i$. The *index sum* of β is $\sum_{i=1}^{n} k_i \cdot |l_i|$ (i.e., each letter, not just syllable in β, contributes to that sum).

For 3-braid words, the k_i interchange between 1 and 2. Thus the vector of the l_i, considered up to cyclic permutations, determines the conjugacy class unambiguously. We call it the *Schreier vector*.

Let β be a positive word. We call $\tilde{\beta}$ an *extension* of β if $\tilde{\beta}$ is obtained by replacing some (possibly no) trivial syllables in β by non-trivial ones of the same index. Contrarily, we call β a *syllable reduction* of $\tilde{\beta}$. We call β *non-singular* if $[\beta]_k > 1$ for all $k = 1, \ldots, n-1$, where $[\beta]_k = \sum_{k_i = k} l_i$ is the exponent sum of σ_k in β. If $[\beta]_k = 1$, we say that the syllable of index k in β is *isolated* or *reducible*.

To avoid confusion, it seems useful to clarify a priori the following use of symbols (even though we recall it at appropriate places later).

A comma separated list of integers will stand for a *sequence of syllable indices* of a braid word. The "at" sign "@", written after such an index means that the corresponding syllable is trivial, while by an exclamation mark "!" we indicate that the syllable is non-trivial. (As the exponent for non-trivial syllables will be immaterial, it is enough to distinguish only whether the syllable is trivial or not.) If none of ! and @ is specified, we do not exclude explicitly any of either types.

A bracketed but non-comma separated list of integers will stand for a braid word. An asterisk "*" put after a letter (number) in such a word means that this letter may be repeated (it need not be repeated, but it must not be omitted). So a (possibly trivial) index-2 syllable can be written as 2^*. The expression "[23]" (within another pair of brackets) should mean a letter which is either "2" or "3", and "[23]+" means a possibly empty sequence of letters "2" and "3". (See pp. 33 and 49 for examples of use.)

2.5 Braid Representations of Links

By a theorem of Alexander [Al], any link is the *closure* $\hat{\beta}$ of a braid β. The *braid index* $b(L)$ of a link L is the smallest number of strands of a braid β whose closure $\hat{\beta}$ is L. See [Mo, FW, Mu]. Such β are also called *braid representations* of L. The closure operation gives for a particular braid word β a link diagram $D = \hat{\beta}$. Then we have, for example, $w(\hat{\beta}) = [\beta]$, $c(\beta) = c(\hat{\beta})$, $c_{\pm}(\beta) = c_{\pm}(\hat{\beta})$, and $s(\hat{\beta})$ is the number of strings of β (i.e., n for $\beta \in B_n$).

Many properties of braid words we will deal with relate to the corresponding properties of their link diagrams. For example, a braid word is called *positive* if it contains no σ_i^{-1}, or in other words, its closure diagram is positive. A braid is *positive* if it has a positive braid word. Likewise can be done with almost positive and negative. In a similar fashion, we say that braid is (*A/B-*)adequate if it has an (*A/B-*)adequate word representation, and a word is adequate if the link diagram obtained by its closure is adequate.

If a braid word β is written as $\sigma_1^{\pm 1}\alpha\sigma_1^p\alpha'$, where $p \in \mathbb{Z}$ and none of α and α' contains a syllable of index 1, then the diagram admits a *flype*, which exchanges the syllables of index 1 in β, so that we obtain $\sigma_1^p\alpha\sigma_1^{\pm 1}\alpha'$. This operation preserves the isotopy type of the closure link $\hat{\beta}$, but in general changes the braid conjugacy class. The phenomenon is explained in [BM]. In the context of general link diagrams, the flype has been studied also extensively, most prominently in [MT].

Alternatively to the standard Artin generators, one considers also a representation of the braid groups by means of an extended set of generators (and their inverses)

$$\sigma_{i,j}^{\pm 1} = \sigma_l \ldots \sigma_{j-2}\sigma_{j-1}^{\pm 1}\sigma_{j-2}^{-1} \ldots \sigma_i^{-1}$$

for $1 \le i < j \le n$. Note that

$$\sigma_i = \sigma_{i,i+1} . \tag{2.10}$$

A representation of a braid β, and its closure link $L = \hat{\beta}$, as word in $\sigma_{i,j}^{\pm 1}$ is called a *band representation* [BKL]. A band representation of β spans naturally a Seifert surface of the link L as in Figure 1.1: one glues disks into the strands, and connects them by half-twisted bands along the $\sigma_{i,j}$. The resulting surface is called *braided Seifert surface* of L. Bennequin proved in [Be]:

Theorem 2.2 (Bennequin) *For a 3-braid link, a minimal genus braided Seifert surface band representation always exists on a 3-braid.*

Followingly, minimal genus Seifert surface of L, which is a braided Seifert surface, is also called a *Bennequin surface* [BM2].

In this monograph we will deal exclusively with band representations in B_3. Then we have three band generators $\sigma_{i,i \bmod 3+1}$ (where $i = 1, 2, 3$, and "mod" is taken with values between 0 and 2). With (2.10), we have $\sigma_1 = \sigma_{1,2}$ and $\sigma_2 = \sigma_{2,3}$, and with the special meaning of $\sigma_3 \in B_3$ introduced above, $\sigma_3 = \bar{\sigma}_{1,3}$, where bar denotes the mirror image. (This mirroring is used here for technical reasons related to Xu's normal form, as explained below.)

If a band representation contains only positively half-twisted bands (i.e., no $\sigma_{i,j}^{-1}$ occur), it is called *(band-)positive* or *strongly quasi-positive*. A link with a strongly quasi-positive band (braid) representation is called strongly quasi-positive. Such links have an importance in connection to algebraic curves; see [Ru2].

If a band representation contains exactly one negative band, we call it *almost strongly quasi-positive*. If it contains k negative bands, we can call it *k-almost strongly quasi-positive*. Similarly we can define a *band-negative* representation and *(k-)almost strongly quasi-negative*, and various versions of *strongly quasi-signed* for braids and links as a common term for the corresponding types of strongly quasi-positive and strongly quasi-negative objects.

Using the skein polynomial, define a quantity by

$$MFW(L) = \frac{1}{2}\left(\max \deg_v P - \min \deg_v P\right) + 1. \tag{2.11}$$

The *Morton-Franks-Williams braid index inequality* [Mo, FW] (abbreviated as MFW) states that

$$b(L) \geq MFW(L) \tag{2.12}$$

for every link L. This inequality is often exact (i.e., an equality). The study of links where it is exact or not has occupied a significant part of previous literature. Most noteworthy is the work of Murasugi [Mu] and Murasugi–Przytycki [MP2].

The Morton-Franks-Williams inequality results from two other inequalities, due to Morton, namely that for a diagram D, we have

$$1 - s(D) + w(D) \leq \min \deg_v P(D) \leq \max \deg_v P(D) \leq s(D) - 1 + w(D). \tag{2.13}$$

Franks–Williams showed these inequalities for the case of braid representations (i.e. when $D = \hat{\beta}$ for some braid β). Later it was observed from the algorithm of Yamada [Y] and Vogel [Vo] that the braid version is actually equivalent to (and not just a special case of) the diagram version.

These inequalities were later improved in [MP2] in a way that allows to settle the braid index problem for many links (see Chapter 6 or also [Oh]).

In the special case of 3-braids (and with the special meaning of σ_3 as described above), Xu [Xu] gives a normal form of a conjugacy class in $\sigma_{1,2,3}$. By Xu's algorithm, each $\beta \in B_3$ can be written in one of the two forms

(A) $[21]^k R$ or $L^{-1}[21]^{-k}$ $(k \geq 0)$, or
(B) $L^{-1} R$,

where L and R are positive words in $\sigma_{1,2,3}$ with (cyclically) non-decreasing indices (i.e., each σ_i is followed by σ_i or $\sigma_{i \bmod 3+1}$, with "mod" taken with values between 0 and 2). Since the form (B) must be cyclically reduced, we may assume that L and R do not start or end with the same letter. This form is the shortest word in $\sigma_{1,2,3}$ of a conjugacy class. By Theorem 2.2, the braided surface is then a minimal genus (or Bennequin) surface. (Remark: Below we will make clear each time from the context whether we use the letter L to denote the compound of Xu's form, or a link.)

2.6   Gauß Sum Invariants

We recall briefly the definition of Gauß sum invariants. They were introduced first in [Fi] for braids, and later [Fi2, PV] for knots. It is known that all they give formulas for Vassiliev invariants.

Definition 2.2 ([Fi2])   A Gauß diagram of a knot diagram is an oriented circle with arrows connecting points on it mapped to a crossing and oriented from the preimage of the undercrossing (underpass) to the preimage of the overcrossing (overpass).

We will call a pair of crossings whose arrows intersect in the Gauß diagram a *linked* pair.

Example 2.1   As an example, Figure 2.3 shows the knot 6_2 in its commonly known projection and the corresponding Gauß diagram.

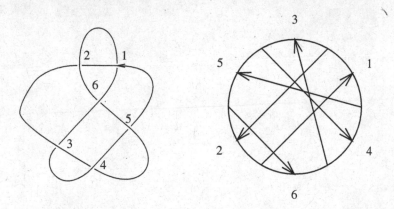

Fig. 2.3   The standard diagram of the knot 6_2 and its Gauß diagram

The simplest (non-trivial) Vassiliev knot invariant is the *Casson invariant* v_2, with $v_2 = \Delta''(1)/2 = -V''(1)/6$, for which Polyak–Viro [PV, PV2] gave the simple Gauß sum formula

$$v_2 = \bigotimes \,. \tag{2.14}$$

Here the point on the circle corresponds to a point on the knot diagram, to be placed arbitrarily except on a crossing. (The expression does not alter with the position of the basepoint; we will hence have, and need, the freedom to place it conveniently.)

We will use the symmetrized version of (2.14) w.r.t. taking the mirror image of the knot diagram:

$$v_2 = \frac{1}{2} \left(\bigotimes + \bigotimes \right) \,. \tag{2.15}$$

Chapter 3
Xu's Form and Seifert Surfaces

We studied the relation of Xu's algorithm and the skein polynomial in [St2], and here we will go further to connect fiberedness, and the Alexander and Jones polynomial to Xu's form.

3.1 Strongly Quasi-Positive Links Among Links of Braid Index 3

Theorem 1.1 follows relatively easily from the work in [St2], but it is a good starting point for the later more substantial arguments basing on Xu's form. It legitimizes the term "strongly quasi-positive 3-braid link," which otherwise would be linguistically sloppy, for it may mean a 3-braid link that is strongly quasi-positive, or the link (which is the closure) of a strongly quasi-positive 3-braid.

Proof of Theorem 1.1 The inequalities of (2.13) for the v-degree of P applied on a positive band representation show that, with $P = P(L)$ and $\chi = \chi(L)$,

$$\min \deg_v P \geq 1 - \chi. \tag{3.1}$$

Because of [LM, Proposition 21], which now says

$$P(v, v^{-1} - v) = 1, \tag{3.2}$$

we have $\min \deg_v P \leq \max \deg_z P$, and by Stoimenow [St2] $\max \deg_z P = 1 - \chi$. So from (3.1) we obtain

$$\min \deg_v P = 1 - \chi. \tag{3.3}$$

© The Author(s) 2017
A. Stoimenow, *Properties of Closed 3-Braids and Braid Representations of Links*,
SpringerBriefs in Mathematics, https://doi.org/10.1007/978-3-319-68149-8_3

It is known that the minimal degree term in z of the skein polynomial of an n-component link is divisible by $(v - v^{-1})^{n-1}$. (In [Kn] in fact all occurring terms are classified.) So if $MFW(K) = 1$, then K is a knot. Now, the identity (3.2) implies that if $MFW = 1$ for some knot K, then $P(K) = 1$. For any non-trivial knot $1 - \chi = 2g > 0$, so for any non-trivial strongly quasi-positive knot from (3.1) we have $\min \deg_v P > 0$, and so $P \neq 1$.

So a strongly quasi-positive link cannot have $P = 1$, and always $MFW \geq 2$. Therefore, when the braid index is 3, we have $MFW \in \{2, 3\}$. Then (3.3) and the inequalities (2.13) for the v-degree of P show that a 3-braid representation β has exponent sum $[\beta] = 3 - \chi$, unless $MFW = 2$ and $[\beta] = 1 - \chi$.

In former case, we can find a minimal genus band representation from β by Xu's algorithm [Xu], and this representation must be positive. In latter case, we will have one negative band, and have Xu's form $L^{-1}R$. Here L and R are positive words in the letters $\sigma_1, \sigma_2, \sigma_3 = \sigma_2\sigma_1\sigma_2^{-1}$ with σ_i followed by σ_i or $\sigma_{i \bmod 3+1}$ (as described in Section 2.5), and L is a single letter.

W.l.o.g. assume $L = \sigma_3$. If the first or last letter in R is a σ_3, then we can cancel two bands, and have a positive representation. If some σ_3 occurs in R, then we can write $L^{-1}R = \sigma_3^{-1}\alpha\sigma_3\alpha'$ where α, α' are positive band words. Such a representation is quasi-positive with a Seifert ribbon [Ru] of smaller genus. However, it is known that for strongly quasi-positive links the genus and 4-genus coincide (see, for example, [St]). If R has no σ_3, then we have up to cyclic letter permutation a braid of the form $\sigma_1^k\sigma_2^l\sigma_1^{-1}\sigma_2^{-1}$ with $k \geq 0, l > 0$. This braid is easily seen to reduce to (a positive one on) two strands. □

Remark 3.1 It is not true that all 3-braids that have the skein polynomial of $(2, n)$-torus links have positive 3-band representations. Birman's [B] construction (see Definition 4.1 below) yields examples like 12_{2037} in [HT] with one negative band. This construction gives in fact all non-obvious 3-braid $MFW = 2$ examples, a circumstance shown in [St4] by applying Theorem 4.5 (and reportedly in previous unpublished work of El-Rifai). The result is explained and used below in the proof of Theorem 7.1.

3.2 Uniqueness of Minimal Genus Seifert Surfaces

As (at least what a priori seems to be) an improvement of Bennequin's theorem 2.2, Birman-Menasco proved in [BM2]:

Theorem 3.1 (Birman-Menasco [BM2]) *For 3-braid links not only some but in fact* any *minimal genus surface is a Bennequin surface.*

Xu used this result and the classification in [BM] to conclude that most 3-braid links have a unique minimal genus surface. Unfortunately, she failed to deal (in the oriented sense) with some exceptions.

The discussion here came about from the desire to complete Xu's uniqueness theorem for Seifert surfaces. While our result will be further improved later using work to follow, we need to introduce some notation and basic tools. We will use the work of Kobayashi [Ko], implying that the property a surface to be a unique minimal genus surface is invariant under Hopf (de)plumbing.

Subsequently Mikami Hirasawa advised me about a subtlety concerning the notion of "uniqueness" which must be explained. Xu's work considers Seifert surfaces unique in the sense isotopic to each other, if we *may move* the link by the isotopy. In particular, such isotopy may interchange link components. However, we demand the isotopy to preserve component orientation. (For unoriented isotopy Xu's result is complete.) Contrarily, Kobayashi's setting assumes uniqueness in the sense that Seifert surfaces are isotopic to each other *fixing* the link. Hirasawa explained that the definitions are not equivalent, and that a unique (if we may move the link) minimal genus surface, may get not unique when one (de)plumbs a single Hopf band. To account for this discrepancy, we should establish proper language.

Definition 3.1 Let us use the term *unique* for Kobayashi's notion of uniqueness (up to isotopy fixing the link), and let us call Xu's notion of uniqueness (up to isotopy which may move the link or permute components, but preserves orientation) *weakly unique*.

Here we state the following extension of Xu's uniqueness theorem.

Theorem 3.2 *Every 3-braid link has a weakly unique minimal genus Seifert surface.*

Proof Birman-Menasco showed (Theorem 3.1) that any minimal genus Seifert surface of a 3-braid link is isotopic to a Bennequin surface. Then, Xu showed that every conjugacy class of 3-braids carries a canonical (up to oriented isotopy) Bennequin surface. The conjugacy classes of 3-braids with given closure link were classified in [BM]. Most links admit a single conjugacy class, and we are done, as in [Xu]. The exceptional cases are easy to deal with, except the "flype admitting" braids

$$\sigma_1^{\pm p}\sigma_2^{\pm q}\sigma_1^{\pm r}\sigma_2^{\pm 1}. \tag{3.4}$$

The flype interchanges $\pm q$ and ± 1, and (in general) gives a different conjugacy class, which differs from the original one by orientation. So we have (as in [Xu]), two Bennequin surfaces isotopic only up to orientation. We will settle this case now also for oriented isotopy, that is, show that these surfaces are isotopic to themselves with the opposite orientation.

Recall that for a fibered link, a minimal genus surface is the same as a fiber surface, and such a surface is unique (Neuwirth-Stallings theorem). Moreover, by work of Gabai [Ga, Ga2, Ga3] and Kobayashi [Ko] the properties of a surface to be a minimal genus surface, a unique minimal genus surface or a fiber surface are

invariant under Murasugi (de)sum with a fiber surface. In particular, this invariance
holds for (de)plumbing a Hopf band (which is understood to be an unknotted
annulus with one full, positive or negative, twist). Now we note that in the cases
in (3.4) where $p = r$, the flype is trivial (i.e., realized by a conjugacy), so that
the surface is weakly unique. However, since Kobayashi's theorem may fail for
weakly unique surfaces, we cannot reduce our surfaces to this case. We will use
Hopf (de)plumbings to recur all cases to fiber surfaces or a 2-full twisted annulus.
Then we understand that our surfaces are unique, and in particular weakly unique.
The type of Hopf (de)plumbings we will apply is to interconvert all powers of a
given band generator of given sign (for example, σ_3^k is equivalent to σ_3^{-1} for each
$k < 0$).

  W.l.o.g. assume in (3.4) that we have $-$ in ± 1 and that none of p, q, r is 0
(the other cases are easy). We assume $p, q, r > 0$ and vary the signs before p, q, r
properly. Also, since the flype interchanges $\pm p$ and $\pm r$, we may assume $\pm p \geq \pm r$.
So we exclude the sign choice $(-p, +r)$.

Case 1. $p, q, r > 0$. We can write (3.4) up to cyclic permutation as
$(\sigma_2^{-1}\sigma_1\sigma_2)^p \sigma_2^{q-1} \sigma_1^r$, and conjugating with $\sigma_2\sigma_1\sigma_2$ we have $\sigma_3^p \sigma_1^{q-1} \sigma_2^r$. Since $p, r > 0$,
this is Hopf plumbing equivalent to $\sigma_3\sigma_1\sigma_2$ or $\sigma_3\sigma_2$. These cases are a disk and a
2-full twisted annulus for the reverse $(2, 4)$-torus link, and we are done.

Case 2. $p, r > 0$, $-q < 0$. We have an alternating braid, which is fibered.

Case 3. $p, q, r < 0$. We have a negative braid, which is fibered.

Case 4. $-p, -r < 0$, $q > 0$. We have a word of the form $\sigma_2^q \sigma_1^{-r} \sigma_2^{-1} \sigma_1^{-p}$. If one of
p and r is 1, then we have a $(2, .)$-torus link. If $q = 1$, then we can go over to the
mirror image, and land in case 1. Now, when $p, q, r \geq 2$, the minimal word in Xu's
form is $[2^{q-2}(-3)(-2)^{r-1}(-1)^{p-1}]$. Since $p - 1$ and $r - 1$ are non-zero, its surface
is plumbing equivalent to the one of $[2 - 3 - 2 - 1]$ (if $q > 2$) or $[-3 - 2 - 1]$ (if
$q = 2$). These are the annulus for the reverse $(2, -4)$-torus link, and the fiber of the
$(-2, -2, 2)$-pretzel link, and we are done.

Case 5. $p > 0$, $-r < 0$, $q > 0$. We have

$$\sigma_1^p \sigma_2^q \sigma_1^{-r} \sigma_2^{-1} = \sigma_1^p \sigma_2^{q-1}(\sigma_2 \sigma_1^{-1} \sigma_2^{-1})^r = \sigma_1^p \sigma_2^{q-1} \sigma_3^{-r}.$$

Such a surface is plumbing equivalent to the one for $\sigma_1\sigma_3^{-1}$ (if $q = 1$) resp. $\sigma_1\sigma_2\sigma_3^{-1}$
(if $q > 1$), which are the fibers of the unknot and Hopf link resp.

Case 6. $p > 0$, $-r < 0$, $-q < 0$. We have

$$\sigma_1^{-r} \sigma_2^{-1} \sigma_1^p \sigma_2^{-q} = \sigma_1^{1-r}(\sigma_1^{-1} \sigma_2^{-1} \sigma_1)\sigma_1^{p-1} \sigma_2^{-q} = \sigma_1^{1-r} \sigma_3^{-1} \sigma_1^{p-1} \sigma_2^{-q}.$$

If $-r = -1$, then we have a $(2, p - q - 1)$-torus link. If $-r < -1$, then the band
surface from the right word above is plumbing equivalent to the one for $\sigma_1^{-1} \sigma_3^{-1} \sigma_2^{-1}$
(if $p = 1$) or $\sigma_1^{-1} \sigma_3^{-1} \sigma_1 \sigma_2^{-1}$ (if $p > 1$), which are the annulus for the reverse $(2, 4)$-
torus link, and the fiber of the $(2, 2, -2)$-pretzel link.   □

3.3 Fiberedness

For the rest of the monograph we normalize Δ so that $\Delta(1) = 1$ and $\Delta(t) = \Delta(1/t)$.

Theorem 3.3 *Let L be a strongly quasi-positive 3-braid link. Then the following are equivalent:*

(1) L's minimal genus surface is a Hopf plumbing,
(2) L's minimal genus surface is a fiber surface,
(3) $\max \deg \Delta(L) = 1 - \chi(L)$ and $\max \operatorname{cf} \Delta(L) = \pm 1$,
(4) L's Xu normal form is not R, with syllable length of R divisible by 3. In other words, the Xu normal form is not an extension (in the sense of Definition 2.1) of $[(123)^k]$ for $k > 0$.
(5) Some minimal band form of L contains $\sigma_1^k \sigma_2^l \sigma_1^m$ or $\sigma_2^k \sigma_1^l \sigma_2^m$ as subword for $k, l, m > 0$, or is $\sigma_1^k \sigma_2^l$.

Proof (1) $\Longrightarrow$ (2) $\Longrightarrow$ (3) are clear.

(5) $\Longrightarrow$ (1). Assume after deplumbing, all letters occur in single power, and up to conjugacy the word starts with $\sigma_1 \sigma_2 \sigma_1$ or $\sigma_2 \sigma_1 \sigma_2$. By adjusting one of the two, we can have a σ_1^2 or σ_2^2 if the next letter is σ_1 or σ_2. In that case we deplumb a Hopf band. If the letter after $\sigma_1 \sigma_2 \sigma_1$ or $\sigma_2 \sigma_1 \sigma_2$ is σ_3, then we have $\sigma_1 \sigma_2 \sigma_1 \sigma_3 = \sigma_1 \sigma_2^2 \sigma_1$, can can also deplumb a Hopf band. Then we reduce the surface for that of $\sigma_1 \sigma_2 \sigma_1$ which is the Hopf band.

(3) $\Longrightarrow$ (4). We prove the contrary. Assume (4) does not hold. Band-positive surfaces are always of minimal genus, so that the properties we investigate are invariant under Hopf (de)plumbings. Under applying skein relations at non-trivial syllables we are left with powers of $\sigma_1 \sigma_2 \sigma_3$. Apply the skein relation for Δ at the last band. Then β_- and β_0 are both of minimal length. We already proved, in (5) $\Rightarrow$ (1), that β_0 is fibered, so $\max \deg \Delta = 1 - \chi(\beta_0)$ and $\max \operatorname{cf} \Delta = \pm 1$. The same holds for β_- by [St2, Proposition 2], since β_- is of Xu's minimal form $L^{-1}R$ and is not positive. So the terms in degree $1 - \chi(\beta)$ of $\Delta(\beta)$ either cancel or give ± 2.

(4) $\Longrightarrow$ (5). We prove the contrary. Assume (5) does not hold. If we do not have a word $\beta = R$ in Xu's form of length divisible by 3, another option would be to have a word with cyclically decreasing indices. But note that $[321] = [121]$ is the Hopf band, while $[321321] = [211211]$ contains a 121, too, and so we are done. Otherwise the index array of the syllables of β must contain the same entry with distance 2, and such a word is conjugate to the ones excluded.   □

Theorem 3.4 *Any closed braid of Xu's form $L^{-1}R$ is fibered.*

The proof uses some work of Hirasawa-Murasugi. A consequence of their result is the following lemma, which we require. It is proved in Appendix A.1.

Lemma 3.1 *The links L_k, given by the closed 3-braids $[(123)^k - 2]$ for $k > 0$, are fibered.*

Proof of Theorem 3.4 We use induction on the length of $L^{-1}R$ and for fixed length on the exponent sum. Under Hopf deplumbings assume all syllables in L and R are trivial. Assume up to mirroring that L is not shorter than R. Permute by conjugacy R to the left, and permute the indices so that L starts with -3.

The following transformations also offer a Hopf deplumbing

$$1 - 3 - 2 - 1 \to -21 - 2 - 1 \to -2 - 2 - 12 \to -2 - 12$$

and

$$12 - 3 \to -211$$

These reductions fail both if either L and R have length at most 2 or R has length 1. (Remember L is not shorter than R.) In former case one checks directly that one has a disk, Hopf band or connected sum of two Hopf bands. We consider latter case.

By conjugacy permute the indices so that $R = 1$; also assume L has length at least 3. If L starts and ends with -2, then we do

$$1 - 2 - 1 \cdots - 2 \to -23 - 1 \cdots - 2 \to 3 - 1 \cdots - 2 - 2$$

(where the right transformation is a conjugacy) and deplumb a Hopf band. If L starts with -3, then we transform as before

$$1 - 3 - 2 - 1 \to -21 - 2 - 1 \to -2 - 2 - 1 - 2$$

and deplumb a Hopf band. So L starts with -2 and ends on -3. Then the mirror image of β is up to conjugacy of the form $[(123)^k - 2] = [(1221)^k(-2)^{k+1}]$, which we dealt with in the lemma before.   □

Combinedly, we obtain

Corollary 3.1 *Let W be a 3-braid link. Then the following are equivalent:*

(1) W is fibered,
(2) $\max \deg \Delta(W) = 1 - \chi(W)$ and $\max \mathrm{cf}\, \Delta(W) = \pm 1$,
(3) W's Xu form is not of the type L^{-1} or R, with syllable length of L or R divisible by 3.   □

Example 3.1 The routine verification, with Mikami Hirasawa, of the tables in [HT] for fibered knots, has shown non-fibered knots with monic Alexander polynomial of degree matching the genus start at 12 crossings. One of these knots, 12_{1752}, has braid index 4, so that (as expected) the corollary does not hold for 4-braids.

Corollary 3.2 *Any non-split closed 3-braid $\hat{\beta}$ with $|[\beta]| \le 2$ is fibered, in particular so is any slice 3-braid knot.*   □

Again 6_1 and 9_{46} show the second part is not true for 4-braids. The first statement in the corollary extends simultaneously the property of amphicheiral knots, which follows also from [BM], since amphicheiral 3-braid knots are closed alternating 3-braids.

Originally the insight about a unique minimal genus surface motivated the fibered 3-braid link classification. Still this insight lacks asset as to the somewhat improper way it emerges. Note, for example, that Birman-Menasco's stronger version in Theorem 3.1 of Bennequin's theorem 2.2 enters decisively into [Xu] and the proof of Theorem 3.2, but then paradoxically latter imply that Birman-Menasco's formulation is actually equivalent to, and not really an improvement of, Bennequin's theorem.

Also, Birman-Menasco's 3-braid link classification [BM] prioritizes Schreier's conjugacy algorithm and fails to lend any geometric understanding to Xu's normal form, while such an interpretation becomes evident in our setting. This provided strong motivation for Theorem 1.2. Its proof is completed by dealing with the non-fibered cases in Appendix A.2. It requires also a part of the further detailed consideration of the Alexander polynomial that follows next.

While we, in the bottom line, still cannot provide an "easy" proof of Birman-Menasco's classification (see Remark on p. 34 of [BM]), our work will likely lead to simplifications in their very lengthy treatment, which makes heavy use of incompressible surfaces.

Chapter 4
Polynomial Invariants

4.1 Alexander Polynomial

In [B], Birman proposed (but considered as very difficult) the problem to classify 3-braid links with given polynomials. In [St2] we dealt with the skein polynomial. Now we can extend our results to the Alexander polynomial (with the convention in the beginning of Section 3.3). The following discussion gives a fairly exact description how to find the 3-braid links, if such exist, for any possible admissible (as specified in Section 2.2) polynomial.

A solution for the Jones polynomial is presented afterwards.

Lemma 4.1 *If β is strongly quasi-positive and fibered, then* $\max \operatorname{cf} \Delta = +1$.

(Here it is essential to work with the leading, not trailing coefficient of Δ and with strongly quasi-positive links and not their mirror images.)

Proof We know $\min \deg_v P = \max \deg_z P = 1 - \chi$. So $[P]_{z^{1-\chi}}$ has a term in degree $v^{1-\chi}$. The coefficient must be $+1$ because of (3.2) and because it is the only coefficient that contributes to the absolute term in $P(v, v^{-1} - v)$. Now from the classification of leading z-terms of P in [St2] it follows that $[P]_{z^{1-\chi}}$ can have at most one further term, with coefficient ± 1. If such term exists, the substitution (2.4) would either cancel the terms in degree $t^{(1-\chi)/2}$ in $\Delta(t)$ or give coefficient ± 2, so our link cannot be fibered. Thus a second term does not exist, and the claim follows from (2.4). □

Lemma 4.2 *Let $\beta' = [(123)^k]$ be an even power k of $[123]$. Assume β has Xu normal form R, and after syllable reduction becomes β', but $\beta \neq \beta'$ itself (i.e., some syllable in β is non-trivial). Then $\hat{\beta}$ satisfies* $\max \deg \Delta = -1 - \chi$; *moreover,* $\max \operatorname{cf} \Delta > 0$ *and is equal to the number of non-trivial syllables in β.*

Proof We use for fixed k induction on the exponent sum. If exactly one syllable is non-trivial with exponent 2, then applying the skein relation at the exponent-2 syllable shows that $\beta = \beta_+$ inherits the Alexander polynomial of β_- with positive

A. Stoimenow, *Properties of Closed 3-Braids and Braid Representations of Links*, SpringerBriefs in Mathematics, https://doi.org/10.1007/978-3-319-68149-8_4

sign (since $\beta_0 = \beta'$, whose closure has zero polynomial). Now β_- is positive and has an index array with a subsequence of the form xyx, and so is fibered by Theorem 3.3. Then by Lemma 4.2, max cf $\Delta = +1$ and max deg $\Delta = 1 - \chi(\hat{\beta}_-) = -1 - \chi(\hat{\beta})$.

If $[\beta] - [(123)^k] > 1$, then applying the skein relation at any non-trivial exponent 2 syllable gives (with positive sign) the Alexander polynomials of two closed braids β_0 and β_-, former of which is fibered and latter of which has the requested property by induction. Then the maximal terms in degree $1 - \chi(\hat{\beta}_-) = -\chi(\hat{\beta}_0)$ are positive and do not cancel. The one of β_- is $+1$, while the one of β_0 by induction one less than the number of non-trivial syllables in β (since in β_0 one more syllable becomes trivial).

If some syllable in β has exponent > 2, then β_- is not minimal, and the degree and leading coefficient of Δ are inherited (with positive sign) from β_0. $\square$

Lemma 4.3 *Assume β has Xu normal form R, which after syllable reduction becomes an odd power k of* [123]. *Then* max deg $\Delta = 1 - \chi$, *and* max cf $\Delta = +2$.

Proof First we prove the claim if $\beta = [(123)^k]$. We know that max deg $\Delta = 1 - \chi$ and max cf $\Delta = \pm 2$, so we must exclude max cf $\Delta = -2$. Applying the skein relation gives the polynomials of β_0 and β_- with positive sign. β_0 is positive and fibered as before, so max cf $\Delta = +1$. Then clearly β_- (which is of the form $L^{-1}R$ and also fibered) cannot have max cf $\Delta = -3$.

If $\beta \neq [(123)^k]$, then it has a non-trivial syllable. Applying the skein relation at a letter in that syllable we find that β_- reduces. So the leading term comes from β_0, and with positive sign. $\square$

Theorem 4.1 *Fix some admissible Alexander polynomial Δ. Then*

(1) If $\Delta = 0$, then the 3-braid links with such polynomial are the split links and the closures of (incl. negative) even powers of [123].

(2) If $\Delta \neq 0$, there are only finitely many 3-braid links with this Δ. They all have $1 - \chi = 2$ max deg Δ or $-1 - \chi = 2$ max deg Δ. In latter case they are up to mirroring strongly quasi-positive.

(3) If max cf $\Delta \leq -2$, then no 3-braid knot or 3-component link has such Alexander polynomial, and any 2-component link is strongly quasi-negative.

(4) If max cf $\Delta \geq +2$, any 3-braid knot or 3-component link with such Alexander polynomial is strongly quasi-positive or a mirror image of it. Any 2-component link is strongly quasi-positive.

(5) If $|$ max cf $\Delta| > 2$, then any 3-braid link with such Alexander polynomial has 2 max deg $\Delta = -1 - \chi$.

Definition 4.1 Let for $\beta \in B_3$ with $6 \mid [\beta]$, the *Birman dual* β^* be defined by $\beta^{-1}\delta^{2[\beta]/3}$, where $\delta = [121]$ and δ^2 generates the center of B_3.

Birman [B] shows that $\hat{\beta}$ and $\hat{\beta}^*$ have the same skein polynomial. This observation relates to our explanation at a couple of places, for example, in Remark 3.1, and also in the below arguments.

Proof of Theorem 4.1 Let us exclude a priori trivial and split links. The claims follow from the discussion of $2 \max \deg \Delta$ and $\max \operatorname{cf} \Delta$ in cases.

We know by Theorem 3.4 (or by [St2, St4], as noted before) that if $\beta \in B_3$ is not (up to mirroring) strongly quasi-positive, then $2 \max \deg \Delta = 1 - \chi$ and $\max \operatorname{cf} \Delta = \pm 1$.

It remains to deal with the Xu form R. The form L is just the mirror image, and mirroring preserves the Alexander polynomial for knots and 3-component links and alters the sign for 2-component links.

If making trivial all syllables in β, the new word β' is not a power of [123], then we proved that $\hat{\beta}$ is fibered, so again $2 \max \deg \Delta = 1 - \chi$ and $\max \operatorname{cf} \Delta = \pm 1$.

If β' is an even power of [123], then by Birman duality we conclude that $\Delta(\hat{\beta}') = 0$. If $\beta \neq \beta'$, then one uses Lemma 4.2. Thus $2 \max \deg \Delta = -1 - \chi$.

If β' is an odd power of [123], then use the observation in Lemma 4.3 (or make one syllable to exponent 4 and use Birman duality) to conclude that $2 \max \deg \Delta = 1 - \chi$ and $\max \operatorname{cf} \Delta = +2$. □

Example 4.1 In certain situations this theorem gives the most rapid test to exclude closed 3-braids. For example, the knot 13_{6149} has $MFW = 3$, but seeing that Δ has $\max \operatorname{cf} \Delta = -2$ we immediately conclude that it cannot be a closed 3-braid.

Corollary 4.1 *There are only finitely many 3-braid links with given* $\max \deg \Delta$ *(provided* $\Delta \neq 0$*).* □

It is actually true (as we will prove below) that for links with *any* bounded braid index there are only finitely many different Alexander polynomials of given degree admitted. However, such Alexander polynomials may be admitted by infinitely many different links (of that braid index). See Section 7.3 for some remarks.

Corollary 4.2 *Non-split 3-braid links bound no disconnected Seifert surfaces (with no closed components). In particular there are no non-trivial 3-braid boundary links.*

Proof For such links the Alexander polynomial is zero. The only non-split 3-braid links L of zero polynomial are closures of $[(123)^k]$ for even k. These cases are easily ruled out by linking numbers. Any pair of components of L has non-zero linking number. So a connected component of any Seifert surface S of L must have at least two boundary components of L, and if S is disconnected, L has at least 4 components, which is clearly not the case. □

The examples that falsify this claim for 4-braids are again easy: consider the (closure links of) words in $[1 \pm 2 - 1]$ and $[2 \pm 3 - 2]$.

Corollary 4.3 *If* Δ *is the Alexander polynomial of a 3-braid link, then* $|\max \operatorname{cf} \Delta| \leq \max \deg \Delta + 2$. *For a knot* $|\max \operatorname{cf} \Delta| \leq \max \deg \Delta + 1$.

Proof If $|\max \operatorname{cf} \Delta| \leq 2$, then we are easily done. ($\Delta = \pm 2$ cannot occur for a knot.) Otherwise we have up to mirroring a strongly quasi-positive word β reducing to an even power of [123]. Now $|\max \operatorname{cf} \Delta|$ counts non-trivial syllables, so $[\beta] = 3 - \chi \geq 2|\max \operatorname{cf} \Delta|$, and $-1 - \chi = 2 \max \deg \Delta$. For a knot the number of syllables with exponent $\neq 2$ is at least 2, so $[\beta] - 2 \geq 2|\max \operatorname{cf} \Delta|$. □

Note in particular that the proof shows that $|\max \mathrm{cf}\,\Delta|$ can be any given natural number, and how to find the link that realizes this number. We emphasize this here, because later we will prove a contrary statement in the case of *alternating* links for *every* arbitrary braid index (see Corollary 6.6).

Corollary 4.4 *A 3-braid* knot *is fibered if and only if* $\max \mathrm{cf}\,\Delta = \pm 1$.

Proof It remains to explain why no 3-braid knot has $\max \mathrm{cf}\,\Delta = \pm 1$ but $\max \deg \Delta < 1 - \chi$. Latter condition would imply that we have up to mirroring a strongly quasi-positive word β reducing to an even power of [123], and former condition that β has only one non-trivial syllable. But $[(123)^{2k}]$ has 3-component closure, and making one syllable non-trivial cannot give a knot. $\square$

In particular, it is worth noting

Corollary 4.5 *No non-trivial 3-braid knot has trivial Alexander polynomial.* $\square$

Again, the two 11 crossing knots immediately show that this is not true for 4-braids.

Example 4.2 We can also easily determine the 3-braid links for some small degree Alexander polynomials. For example, we see that no other 3-braid knot has the polynomial of 3_1, 4_1, or 5_2. Similarly we can check that no 3-braid knot has the polynomial of 9_{42} and 9_{49} (which shows that these knots have braid index 4), a fact we will derive in the last chapter using entirely different representation theory arguments.

4.2 Jones Polynomial

The control of the Jones polynomial on 3-braid links was the object of main attention in [B]. We can accomplish this with a similar argument to Δ. The result we obtain can be conveniently described in our setting and is as follows:

Theorem 4.2 *Let L be a non-split 3-braid link, and* $L = \hat{\beta}$ *with* $\beta \in B_3$. *Then*

$$\mathrm{span}\,V(L) \le 4 - \chi(L).\tag{4.1}$$

Equality holds if and only if L is strongly quasi-signed (i.e. -positive or -negative, or equivalently $\|[\beta]\| = 3 - \chi(L)$*), and not fibered. More specifically, the following holds:*

(1) If L is strongly quasi-positive, then $\quad \min \deg V = \dfrac{1 - \chi}{2}$ *and* $\quad \min \mathrm{cf}\,V = \pm 1$.

Analogously, if L is strongly quasi-negative, then $\quad \max \deg V = \dfrac{\chi - 1}{2}$ *and* $\max \mathrm{cf}\,V = \pm 1$.

(2) If L is strongly quasi-positive and fibered, then $\operatorname{span} V(L) \leq 3 - \chi(L)$. *If L is strongly quasi-positive and not fibered, then (4.1) is an equality and* $\max \operatorname{cf} V = \pm 1$. *(The properties for strongly quasi-negative are analogous.)*

(3) If L is not strongly quasi-signed and $\|[\beta]\| < 1 - \chi(L)$, *then*

$$\min \operatorname{cf} V(L) = \pm 1, \quad \max \operatorname{cf} V(L) = \pm 1 \quad \text{and} \quad \operatorname{span} V(L) = 3 - \chi(L).$$

(4) If L is not strongly quasi-signed and $[\beta] = 1 - \chi(L)$, *then*

$$\max \operatorname{cf} V(L) = \pm 1 \quad \text{and} \quad \max \deg V(L) = \frac{5}{2} - \frac{3}{2}\chi(L).$$

Moreover, $\min \deg V \geq \dfrac{-1 - \chi}{2}$, *and if equality holds, then* $\min \operatorname{cf} V = \pm 2$. *(The case* $[\beta] = \chi(L) - 1$ *is analogous.)*

Apart from solving Birman's problem how to determine 3-braids with given Jones polynomial, Theorem 4.2 easily implies that no non-trivial 3-braid link has trivial (i.e., unlink) polynomial. We defer the discussion of the non-triviality of the Jones polynomial to [St4], where we work in the much more general context of semiadequate links. In that paper we will show that semiadequate links have non-trivial Jones polynomial. This result in fact motivated Theorem 4.5, which then provides a different conclusion about the non-triviality of the polynomial. A further application will be the classification of the 3-braid links with unsharp Morton-Franks-Williams inequality (mentioned in Remark 3.1). Thus (a part of) [St4] is logically a sequel to the present work, even if (for this sequel part) already published.

Corollary 4.6 *For a given Jones polynomial V (actually the pair* $(\min \deg V, \max \deg V)$ *is enough), there are at most three values of* $\chi(L)$ *of a 3-braid link L with* $V(L) = V$. *If* $\min \deg V \cdot \max \deg V \leq 0$, *then* $\chi(L)$ *is unique.*

Proof The theorem shows that the value of χ is determined by one of $\max \deg V$, $\min \deg V$ or $\operatorname{span} V = \max \deg V - \min \deg V$. In particular for a pair $(\min \deg V, \max \deg V)$ there exist at most three values of χ of 3-braid links with a Jones polynomial realizing this pair. If $\min \deg V \cdot \max \deg V \leq 0$, then the options that a 3-braid β with $V(\hat{\beta}) = V$ is strongly quasi-signed or almost quasi-signed are excluded (up to a few simple cases that can be checked directly), so $\chi(L)$ is determined (unambiguously) by $\operatorname{span} V$.   □

In particular, since 3-braid links of given χ are only finitely many, we have

Corollary 4.7 *There are only finitely many closed 3-braids with the same Jones polynomial, actually with the same pair* $(\min \deg V, \max \deg V)$. □

For example, one easily sees that no 3-braid knot has the polynomial of 9_{42}. Similarly, no other 3-braid knot has the polynomial of the figure-8-knot (there is, however, a 4-braid knot with such polynomial, 11_{386}).

That there are only finitely many closed 3-braids with the same skein polynomial was known from [St2]. For the Jones polynomial one should note that infinite families were constructed by Traczyk [Tr] if one allows polynomials up to units. Traczyk's examples show that for fibered (strongly quasi-)positive links span$V$ may remain the same while $\chi \to -\infty$, so one cannot expect a full (lower) control on span$V$ from χ. From these links, one obtains by connected sum infinite families with the same polynomial for 5-braids. (The status of 4-braids remains unclear.) Also Kanenobu [K2] constructed finite families of 3-braids of any arbitrary size, so that our result is the maximal possible.

A question that surfaces naturally with these remarks in mind is

Question 4.1 Does for 3-braid links the Jones polynomial (or Alexander polynomial) determine the skein polynomial? In other words, do any two 3-braid links with the same V (or Δ) have also equal P?

We have at least the following partial result, whose proof we postpone after the proof of Theorem 4.2. See also the comment in the preface of the book.

Corollary 4.8 *A given Jones polynomial* $V = V(L)$ *is realized for 3-braid links* L *by at most three different skein polynomials* $P(L)$. *If* min deg $V \cdot$ max deg $V \leq 0$, *then* $V(L)$ *determines* $P(L)$.

In the general case one cannot expect a positive answer to the above question. At least for links there is now the method of [EKT] available, which should yield large families of links with the same Jones but different skein (or Alexander) polynomial. For constructing families with the same Alexander (but different Jones or skein) polynomial, further techniques are available, applicable also for knots, like the non-faithfulness of the Burau representation (see [Bi]) or tangle surgeries (see [Bl]).

For the proof of Theorem 4.2 we use the previous work in [St2] (that in particular answered Birman's question) on the skein polynomial. We apply again the result in [St2] that max deg$_z P = 1 - \chi$ for closed 3-braids. As in that paper, we distinguish the cases of band-positive, band-negative 3-braids, and such of Xu's form $L^{-1}R$.

By the Morton-Franks-Williams inequalities (2.13), a 3-braid β of exponent sum (writhe) $[\beta]$ has v-degrees of P in $[\beta] - 2$ we call *left degree*, $[\beta]$ we call *middle degree* and $[\beta] + 2$ we call *right degree*. The terms of $[P]_{z^k}$ for some k in these degrees will be called left, middle, and right terms.[1]

Lemma 4.4

(a) *If* β *is band-positive, then* max cf $_z P$ *has* v-*terms in the left degree and possibly in the middle degree. If* β *is band-negative,* max cf $_z P$ *has* v-*terms in the right degree and possibly in the middle degree. In either situation a term in the middle degree occurs if and only if* $\hat{\beta}$ *is not fibered.*

(b) *If* β *is* $L^{-1}R$, *then* max cf $_z P$ *has* v-*terms in the middle degree only.*

In both cases all non-zero occurring coefficients are ± 1.

Proof In (a) we prove only the first claim (the second claim is analogous). Let $\chi = \chi(\hat{\beta})$. Since min deg$_v P \leq$ max deg$_z P = 1 - \chi$ and $1 - \chi = [\beta] - 2 \geq$ min deg$_v P$

---

[1]Note that here the brackets for polynomials and for braids have a completely different meaning.

by MFW, we have min $\deg_v P = \max \deg_z P$, and then (3.2) implies that $\max \mathrm{cf}_z P$ has a v-term in the left degree, with coefficient ± 1. From [St2, Theorem 3] we have then that it has no right-degree term, and that if it has a middle-degree term, the coefficient is ± 1. Now the previous work and the substitution $v = 1, z = t^{1/2} - t^{-1/2}$ for Δ in (2.4) easily show that the middle term occurs if and only if $\hat{\beta}$ is not fibered.

Now consider (b). If $\max \mathrm{cf}_z P$ has a left-degree term, then using the substitution (3.2), we saw that β is band-positive. Otherwise, $[\beta] < 3 - \chi$, so the contribution of the coefficient of $z^{1-\chi} v^{[\beta]-2}$ in (3.2), which is not cancelled, is not in degree 0. Analogously one argues if $\max \mathrm{cf}_z P$ has a right-degree term. $\square$

Note that keeping track of $[\beta]$ and distinguishing between left and middle term is important here. By Remark 3.1, we have links with equal polynomials such that the left term of the one is the middle term of the other.

Lemma 4.5

(a) *If β is $L^{-1} R$ and L has exponent sum $[L] > 1$, then $[P]_{z^{-1-\chi}}$ has a left-degree term, and the coefficient is ± 1.*

(b) *If $[L] = 1$ and $[P]_{z^{-1-\chi}}$ has a left-degree term, then the coefficient is ± 2.*

The two analogous statements hold replacing L by R and left-degree term by right-degree term.

Proof We proved already that $\max \mathrm{cf}_z P$ has only a middle-degree term. So it must remain under the substitution of (3.2) if and only if $[\beta] = 1 - \chi$. The cases (a) and (b) occur when this term must be cancelled, and complemented to 1, respectively.

$\square$

Proof of Theorem 4.2 The statements follow mainly by putting together the last two lemmas and looking in which degrees the non-cancelling contributions of the coefficients of P occur under the substitution (2.5).

If part (a) of Lemma 4.4 applies, we established already (in Theorem 1.1) that the positive band form of β is equivalent to the strong quasi-positivity of $\hat{\beta}$. In this case min $\deg V$ comes from the left-degree term in $\max \deg_z P$. If $\hat{\beta}$ is not fibered, then $[P]_{z^{-1-\chi}}$ must have a right-degree term (with coefficient ± 1) to cancel the middle-degree term of $\max \deg_z P$ under (3.2).

If part (b) of Lemma 4.4 applies, we use the further information of Lemma 4.5. In case (a) of Lemma 4.5, the left and right terms in $[P]_{z^{-1-\chi}}$ determine the degrees and edge coefficients of V.

In case (b) of Lemma 4.5, the maximal term in V comes from the right-degree term in $[P]_{z^{-1-\chi}}$. A coefficient in $t^{(-1-\chi)/2}$ may come only from a left-degree term in $[P]_{z^{-1-\chi}}$, which, if occurring, is with coefficient ± 2. $\square$

This proof underscores the significance of (2.5) as a tool for studying the Jones polynomial. So far it seems to have been useful just for calculating specific Jones polynomials from P. In Chapter 6 we will see further results that come out of considering this substitution.

We note the following equalities that follow from the proof of Theorem 4.2. These will be needed in the study of the Q polynomial, and are also helpful for Corollary 4.8.

Lemma 4.6 *If β is a not strongly quasi-signed 3-braid of exponent sum e and V, χ the Jones polynomial resp. Euler characteristic of its closure, then*

$$\max \deg V - e = \frac{1 - \chi}{2} + 1 \qquad (4.2)$$

$$\min \deg V - e = -\frac{1 - \chi}{2} - 1 \qquad (4.3)$$

Exceptions *occur in* (4.3) *if β is almost strongly quasi-positive, and* (4.2) *if β is almost strongly quasi-negative.*   □

Proof of Corollary 4.8 Depending on whether a 3-braid representation β is strongly quasi-signed, almost quasi-signed, or none of both, V determines $e = [\beta]$ via $\chi(\hat{\beta})$ and/or (4.2) or (4.3). With V and e, one can recover the trace of the Burau matrix $\psi_2(\beta)$, and from that also $P(\hat{\beta})$ (see [B] or (7.15) below). Again, if $\min \deg V \cdot \max \deg V \le 0$, then the options that β is strongly quasi-signed or almost quasi-signed are ruled out easily.   □

4.3 Q Polynomial

We extend the scope of the previous results to the Brandt-Lickorish-Millett-Ho polynomial. The main aim here is to prove

Theorem 4.3 *Only finitely many non-trivial 3-braid links have given Q polynomial, and none has trivial (i.e., unlink) polynomial.*

Note that by Kanenobu's work [K3], again there are finite families of arbitrary large size, so we claim again a sort of contrary result.

While this result may be considered less relevant than its analogs for Δ and V, its proof displays the largest variety of tools necessary to apply, and shows how the various approaches to link invariants (skein relations, state models, and representation theory), which are often considered in isolation, can usefully complement each other. Indeed, Q seems in general more difficult to treat than V. Apart from Kidwell's results [Ki] for alternating links, and those in [St7] for positive *knots*, neither the non-triviality nor the finiteness property seems to have been known previously for any other class of links.

The proof makes use of the full extent of the study of Xu's form. Various additional ingredients will be necessary. One such is J. Murakami's Q-V formula, found in [Mr] (by representation theory, and proved also later by Kanenobu [K] using skein relations). It allows to recur in many cases the problem from Q to V. When dealing with V, we use beside the previous discussion the (Kauffman bracket based) formulas of Section 2.3.

In cases Murakami's formula is not helpful, we apply the Polyak-Viro formula (2.15) for Casson's knot invariant $v_2 = \frac{1}{2}\Delta''(1)$, and the following formula, (likewise skein theoretic and) due to Kanenobu [K4], that relates v_2 to Q.

Theorem 4.4 ([K4]) *For a link L with n components K_i, $i = 1, \ldots, n$, we have*

$$Q'_L(-2) = 3(-2)^n \sum_i v_2(K_i) + 3(-2)^{n-2} \sum_{i<j} lk(K_i, K_j)^2 + (n-1)(-2)^{n-3}, \qquad (4.4)$$

where lk is the linking number.

Using this formula, we prove first a special case of Theorem 4.3.

Proposition 4.1 *Only finitely many (non-trivial) strongly quasi-positive 3-braid links have given Q polynomial, and none has trivial polynomial.*

For the case of knots, we have the following more specific statement, that has also some independent meaning.

Proposition 4.2 *If K is a strongly quasi-positive 3-braid knot, then $v_2(K) \geq g(K)$.*

Note that in [St7] we proved the same inequality for a general *positive* knot. Since the pretzel knots with trivial Alexander polynomial are strongly quasi-positive, nothing like this holds for a general strongly quasi-positive knot, though.

We need some preparations. Let us first fix a convention. We number braid strands by 1, 2, 3 from left to right in the bottom of the braid by and compose words from bottom to top. We propagate strand number through the crossings. In particular strand(number)s may appear permuted in the middle or on top of the braid. To refer to the ordering of the strands with regard to the local position in the braid diagram, we speak of left, middle, and right strand. So, for example, if strands 1 and 2 enter below a σ_1 (strand 1 as left strand and strand 2 as middle strand), then they exit above in the order 2 1 (strand 2 as left strand and strand 1 as middle strand).

Let for a (not necessarily pure, and not closed) braid word the linking numbers of strand i, j numbered as explained above be the sum of the writhes of all crossings these two strands pass. (Note that for links this differs by the additional factor 1/2, which is not relevant, though.)

Lemma 4.7 *Let β be a strongly quasi-positive 3-braid word (not necessarily pure, and not closed). Then the linking numbers l_{ij} between pairs (i, j) of strands satisfy $l_{13} + l_{23} > 0$, unless β is split.*

Proof The sum of writhes of crossings a strand passes through each of $\sigma_{1,2,3}$ is non-negative. So $l_{13} + l_{23} \geq 0$. If $l_{13} + l_{23} = 0$, then strand 3 passes only from below as left or middle strand into a $\sigma_3 = \sigma_1^{-1}\sigma_2\sigma_1$. But since it starts (at the bottom of the braid) as a rightmost strand, this means that it passes no crossing, so β is split. $\square$

We prove Proposition 4.2 by induction on the length of the positive word in $\sigma_{1,2,3}$. Assume β is written as such a word. We can w.l.o.g. cyclically permute the indices.

Lemma 4.8 *If $\hat{\beta}$ is a strongly quasi-positive 3-braid, then a positive word β can be chosen so that it contains a non-trivial syllable, unless β is a power of $[123]$. In particular, it is always possible if $\hat{\beta}$ is a knot, or a fibered link.*

Proof Choose Xu's form. It is $[12]^k \beta'$, where $k \geq 0$ and β' has cyclically non-decreasing indices. This word β has a non-trivial syllable up to cyclic permutations of the letters, unless β' is a power of $[123]$. In this case, if $k > 0$, we apply a YB relation at the initial 121 in β, and are done.   □

Proof of Proposition 4.2 Let $\hat{\beta}$ be a knot. We can w.l.o.g. now assume β is written as a word with a non-trivial syllable s. Now we apply the skein relation of Δ at a letter of s. Let $K = K, K_-, K_0$ be the skein triple, and $\beta_{\pm,0}$ the corresponding braids. Also we can cyclically permute the indices in $\beta_{\pm,0}$. We choose them so that the strand fixed by the permutation of β_0 is number 3. We know from the skein relation of Δ that $v_2(K) - v_2(K_-) = lk(L, M)$ where L, M are the components of K_0. Now that linking number is positive by Lemma 4.7 (unless β_0 is a split braid, but this is impossible because s is non-trivial). By induction on $g(K)$, the claim of Proposition 4.2 follows.   □

Proof of Proposition 4.1 We distinguish three cases depending on the number of components of $\hat{\beta}$.

Case 1. 1 component. In this case $Q'(-2)$, by Kanenobu's formula (4.4), is a multiple of Casson's invariant v_2. The claim we wish to show then is implied by the estimate in Proposition 4.2.

Case 2. 2 components. In this case Kanenobu's formula involves only the square of the linking number of both components K, L of the closure link of β and v_2 of the 2-string subbraid component K. Looking at Xu's normal form, we see that if β is a strongly quasi-positive 3-braid, then writing β as a word in $\sigma_{1,2,3}$, and then expanding $\sigma_3 = \sigma_1^{-1}\sigma_2\sigma_1$, we obtain a word in $\sigma_{1,2}$ of some length c with at most $c/5$ negative letters. Since thus β has at most $c/5$ negative crossings, it is easy to see that not both $v_2(K)$ and $lk(K, L)$ can be zero, and one grows unboundedly when c grows. So we are done.

Case 3. 3 components. In this case Kanenobu's formula (4.4) reduces to the square sum of the linking number of pairs components. Let l_{12}, l_{13}, l_{23} be these linking numbers. Again it is clear that not all of $l_{12}, l_{13}, l_{23} = 0$, and that when $l_{12} + l_{13} + l_{23} \geq 2c/5$ grows unboundedly, then so will $l_{12}^2 + l_{13}^2 + l_{23}^2$, so again we are done.   □

Now we move on to settle the remaining cases in Theorem 4.3. Here the tools differ considerably. Beside the study of Xu's form, we need the help of the following theorem, proved in [St4].

Theorem 4.5 *Any 3-braid is up to conjugacy A-adequate or B-adequate.*

We proved also the following

Theorem 4.6 *A 3-braid word is B-adequate if and only if it is (a) a negative word, or (b) it contains no [121] as subword, and negative entries in the Schreier vector appear isolated. (That is, in cyclic order they are preceded and followed by a positive entry.)*

Using this criterion we prove now

Lemma 4.9 *Assume β is strongly 1- or 2-almost positive and of exponent sum $e \geq 0$. Then β is B-adequate.*

Proof If β is almost strongly quasi-positive, then it is of Xu's form $L^{-1}R$ where L is a single letter. By direct observation we can verify that when permuting indices cyclicly properly and writing out σ_3 in $\sigma_{1,2}$, then the resulting word has no [121] subword, and negative entries in the Schreier vector appear isolated. (In case of $[(312)^k 31 - 2]$, we must cancel 2 crossings first.)

If β is 2-almost strongly quasi-positive, then it is of Xu's form $L^{-1}R$ where L has 2 letters. We can assume that $L = [12]$ or $L = [1^2]$, then we write down the subword of $L^{-1}R$ consisting of the first and last letters of R and L (herein $k \geq 1$):

$$[[23]\ldots 33 - 2 - 1]\quad [[23]\ldots 22 - 1 - 1]$$

$$[[23]\ldots 23 - 2 - 1]\quad [[23]\ldots 31^k 2 - 1 - 1]$$

$$[[23]\ldots 31^k - 2 - 1]\quad [[23]\ldots 33 - 1 - 1]$$

$$[[23]\ldots 23 - 1 - 1]$$

(Here "[23]" means as in Section 2.4 that the word is to begin with "2" or "3".) The reductions and B-adequacy test are done case-by-case. $\square$

Using the formulas in Section 2.3, we prove

Lemma 4.10 *If $L = \hat{\beta}$ is a link, which is the closure of an A-semiadequate 3-braid β, and $V_0 V_1 = -1$, then $V_0 V_2 = 1$ or 2.*

Proof By [St4] an A-semiadequate 3 braid is either positive, in which case $V_1 = 0$, or has no $[-1-2-1]$ and positive entries are isolated in the Schreier vector. In [St4] we described the words for latter type which have $V_1 = 0$. They are (up to braid relations and cyclic letter permutations) of the form[2]

$$[1^{-2}2^{-2}1^{-2}\ldots 1^{-2}2^{-1}1^p 2^{-1}].$$

Analogously to that study, we can see that if $V_0 V_1 = -1$, then β has a word which is obtained from the above type by replacing some 1^{-2} by 1^{-k} for $k \geq 3$. The intertwining graph of the A-state is a path (all vertices have valence 2, except two of valence 1), so $\chi(IG) = 1$. We have $\Delta = 1$ if $k = 3$, and otherwise $\Delta = 0$. So by Theorem 2.1, we have $V_0 V_2 = 1$ for $k = 3$, and $V_0 V_2 = 2$ otherwise. $\square$

To explain what such a property has to do with Q, now we must introduce J. Murakami's formula [Mr]. (See also Kanenobu [K, Theorem 2].)

Let $i = \sqrt{-1}$, $u = \sqrt{-t}$ and $x = u + u^{-1}$. Let further[3] for a braid β of exponent sum e,

$$\chi(\beta, u) = i^e u^{-2e} V_{\hat{\beta}}(t) + u^{-e}(x^2 - 2).\qquad (4.5)$$

Then Murakami's formula is

---

[2]Note that these are exactly the 3-braids which are reducible in the terminology of dynamic properties; this reducibility has nothing to do, though, with the reducibility in Markov's theorem.

[3]Note that V is here what is written as J, and not V, in [K].

Theorem 4.7 (J. Murakami) *If L is a closure of a 3-braid β of exponent sum e, then*

$$Q(L,x) = \chi(\beta, \sqrt{u})^2 - 1 + \frac{2(x^2 + x - 1)}{x^2(x^2 - 3)}(u^e + u^{-e}) + \qquad (4.6)$$

$$+ \frac{-x^4 - 2x^3 + 3x^2 + 4x - 4}{x^2(x^2 - 3)}\chi(\beta, u).$$

Proof of Theorem 4.3 We can w.l.o.g. (taking the mirror image) assume that $e \geq 0$, and (excluding trivial to check special cases) that $\chi \leq 0$. Clearing denominators and absolute terms in (4.6), we find

$$Q_1(L,x) := x^2(x^2 - 3)(Q(L,x) + 1) = \left[i^e(-t)^{-e}V(t) + (-t)^{-e/2}(-t - \frac{1}{t}) \right] \times \quad (4.7)$$

$$\times(-x^4 - 2x^3 + 3x^2 + 4x - 4) +$$

$$+ \left[i^e(-t)^{-e/2}V(-\sqrt{-t}) + (-t)^{-e/4}x \right]^2 \cdot x^2(x^2 - 3) +$$

$$+ 2(x^2 + x - 1)(\sqrt{-t}^e + \sqrt{-t}^{-e}).$$

Rearranging, we need to show that the sum of the following five expressions, regarded as a polynomial in $t^{\pm 1/2}$, has either arbitrarily small minimal degree or arbitrarily large maximal degree. (Note that by (4.6) this sum must be self-conjugate in t up to coefficient signs, which is not at all evident directly.)

$$T_1' = i^e(-t)^{-e}V(t) \cdot (-x^4 - 2x^3 + 3x^2 + 4x - 4)$$

$$T_2' = t^{-e}[V(-\sqrt{-t})]^2 \cdot x^2(x^2 - 3)$$

$$T_3' = 2i^e(-t)^{-3e/4}V(-\sqrt{-t})x^3(x^2 - 3)$$

$$T_4' = (-t)^{-e/2}[(x^2 - 2)(-x^4 - 2x^3 + 3x^2 + 4x - 4) + x^4(x^2 - 3) + 2(x^2 + x - 1)]$$

$$T_5' = (-t)^{e/2}[2(x^2 + x - 1)]$$

Switching $-t \to t$ to simplify the expressions, and regrouping terms, we have

$$T_1 = i^e t^{-e}V(-t) \cdot (-1\ -2\ -1\ -2\ [-4]\ -2\ -1\ -2\ -1)$$

$$T_2 = (-t)^{-e}[V(-\sqrt{t})]^2 \cdot (1\ 0\ 1\ 0\ [0]\ 0\ 1\ 0\ 1)$$

$$T_3 = 2i^e t^{-3e/4}V(-\sqrt{t}) \cdot (1\ 0\ 2\ 0\ 1\ [0]\ 1\ 0\ 2\ 0\ 1)$$

$$T_4 = t^{-e/2}[-2(x - 1)(x^2 - 3)(x^2 - 1)]$$

$$T_5 = t^{e/2}[2(x^2 + x - 1)]$$

Now $u = t^{1/2}$ and $x = t^{1/2} + t^{-1/2}$. The last factors for $T_{1,2,3}$ are given as a list of coefficients, with the absolute term put in brackets. Note that these are polynomials in $\sqrt{t}$, so for example, the first two polynomials have degree 2 in t.

Since we are concerned with cancellations of the leading and trailing coefficients of T_k, let us compile their minimal $m_k = \min \deg_t T_k$ and maximal degrees $M_k = \max \deg_t T_k$ (note that by taking degrees w.r.t. t, and not $\sqrt{t}$, the m_k, M_k are only half-integers).

k	m_k	M_k
1	$\min \deg V - e - 2$	$\max \deg V - e + 2$
2	$\min \deg V - e - 2$	$\max \deg V - e + 2$
3	$-\dfrac{3e}{4} + \dfrac{\min \deg V}{2} - \dfrac{5}{2}$	$-\dfrac{3e}{4} + \dfrac{\max \deg V}{2} + \dfrac{5}{2}$
4	$-\dfrac{e}{2} - \dfrac{5}{2}$	$-\dfrac{e}{2} + \dfrac{5}{2}$
5	$\dfrac{e}{2} - 1$	$\dfrac{e}{2} + 1$

(Here again $\min \deg V = \min \deg_t V(t)$ is a half-integer, and similarly $\max \deg V$.)

Our aim will be to determine what of the M_k is the largest, and to show that the coefficients of the contributing T_k do not cancel. An important case where we exclude (problematic) cancellation is

Lemma 4.11 *Assume that*

$$M_1 = M_2 \geq M_k + 3/2 \quad and \tag{4.8}$$

$$m_1 = m_2 \leq m_k - 3/2 \tag{4.9}$$

for $k = 3, 4, 5$. Then $\max \deg_t Q_1(L, x) \geq M_1 - 1$. (Here $Q_1(L, x)$ refers to (4.7).) If L is the closure of a 3-braid which is A-adequate resp. B-adequate, then assuming (4.9) resp. (4.8) alone is sufficient.

Proof By Theorem 4.5, L is the closure of a 3-braid β which is either A-adequate or B-adequate. We consider the B-adequate case and show that $M_1 = M_2 \geq M_k + 3/2$ implies $\max \deg_t Q_1(L, x) \geq M_1 - 1$. If β is A-adequate, we obtain similarly $\min \deg_t Q_1(L, x) \leq m_1 + 1$, and the result follows by the (anti)symmetry of Q_1.

Now if $M_1 = M_2 \geq M_k + 3/2$, the last three coefficients of $M_{1,2}$ are not cancelled from other M_k (remember these are polynomials in $\sqrt{t}$, and coefficients are meant for such polynomials), so it is enough to show that they do not (completely) cancel among each other. Now a look at the formulas for $T_{1,2}$ shows that if these three coefficients are all to cancel, then (apart from proper sign coincidences) we must have $\bar{V}_1 \bar{V}_0 = -1$ and $\bar{V}_2 = 0$. (Take also into account the switch $-t \to t$.) However, this situation was ruled out by Lemma 4.10.   □

We assumed $e \geq 0$; we also excluded the case $e = 3 - \chi$ of strongly quasi-positive braids. To apply the lemma, we need to establish (4.9) and/or (4.8). For this we use now Lemma 4.6.

Clearly, an almost strongly quasi-negative braid should not be considered for $e \geq 0$; the almost strongly quasi-positive braids require a small extra argument, which is given below. Now

$$\max \deg V - e = \frac{1-\chi}{2} + 1 \geq \frac{e}{2} + 1 , \qquad (4.10)$$

so $M_3 \geq M_4$, and we can discard M_4. Also from (4.10) we obtain

$$M_1 - M_5 - 1 = \max \deg V - \frac{3e}{2} \geq 1 ,$$

so $M_1 > M_5 + 1$, and can neglect M_5. Similarly, if β is not almost strongly quasi-positive, we have with (4.3) and min $\deg V < e/2$ that

$$m_3 < m_4 < m_5 , \qquad (4.11)$$

so $m_{4,5}$ are also irrelevant.

So we need to deal only with m_k, M_k for $k \leq 3$. Moreover, we see from (4.2) that $M_1 \geq (1-\chi)/2 + 3$. So whenever we can apply Lemma 4.11, we have for $\chi \leq 0$ that $\max \deg Q_1 \geq M_1 - 1 > 2$, so $\max \deg_z Q > 0$, and Q is not an unlink polynomial. Moreover, we have for a sequence of links from $\chi \to -\infty$ also $\max \deg M_1 \to \infty$, so $\max \deg Q \to \infty$, and a given (even just degree of a) Q polynomial occurs only finitely many times, as desired. A similar argument applies if we use (4.3).

To apply Lemma 4.11, we need to check $M_3 < M_1 - 1$ and $m_3 > m_1 + 1$. Note that if β is almost strongly quasi-positive, then (4.3) may not hold. So we may not be able to apply Lemma 4.11 directly. However, we remedied this by showing in Lemma 4.9 that β is B-adequate. Then (4.2) holds, and it is enough to use $M_3 < M_1 - 1$ (the condition (4.11) becomes also irrelevant).

Now using (4.2) and (4.3)

$$m_1 - m_3 + 1 = \min \deg V - e - 2 + \frac{3e}{4} - \frac{\min \deg V}{2} + \frac{5}{2} + 1$$

$$= \frac{\min \deg V}{2} - \frac{e}{4} + \frac{3}{2} = \frac{e}{2} - \frac{1-\chi}{4} - \frac{1}{2} \cdot \frac{e}{4} + \frac{3}{2} = \frac{e}{4} - \frac{1-\chi}{4} + 1 ,$$

$$M_1 - M_3 - 1 = \frac{\max \deg V}{2} - \frac{e}{4} - \frac{3}{2} \qquad (4.12)$$

$$= \frac{e}{2} + \frac{1-\chi}{4} + \frac{1}{2} - \frac{e}{4} - \frac{3}{2} = \frac{1-\chi}{4} + \frac{e}{4} - 1 .$$

Then (4.8) becomes equivalent to

$$\frac{1-\chi}{4} + \frac{e}{4} > 1.$$

So, assuming $\chi \le 0$ and $e \le 1 - \chi$, for (4.8) we are left to consider only $(e, \chi) = (0, -1), (0, -3), (1, 0), (1, -2),$ or $(2, -1)$, which are trivial cases to check.

For (4.9) we must deal with *A*-adequate but not *B*-adequate braids with $e \ge -3 - \chi$, i.e. $\le$ 3-almost strongly quasi-positive braids. For $\le$ 2-almost strongly quasi-positive braids we can use Lemma 4.9. While one may extend the argument there to 3-almost strongly quasi-positive braids, it would require a longer case-by-case analysis. We provide instead a different argument.

We try to modify the argument for Proposition 4.1. In case $\hat{\beta}$ is a link of 2 or 3 components, we can argue again as there, and are left with the cases $e = 0, 1$ of 6 and 7 bands, which are easy to deal with.

So assume $\hat{\beta}$ is a knot. Now from (4.12), we have $m_3 = m_1 + 1$, so if still some of the first two coefficients of T_1, T_2 do not cancel, we have min deg$_t Q_1 \le m_1 + 1/2$, and are done. Otherwise again $V_1 V_0 = -1$. So by the proof of Lemma 4.10,

$$\beta = [1^{-k}2^{-2}1^{-2} \ldots 1^{-2}2^{-1}1^l 2^{-1}],$$

with $k \ge 3$ and $l \ge 1$. Since $\hat{\beta}$ is a knot, k and l are easily observed to be odd. Now a $\le$ 3-almost positive word in Xu's form can be chosen, when written out as a word in $\sigma_{1,2}$ of c letters, to have $c_- \le c/5 + 3$ negative letters/crossings. If we reduce the word by cancelling crossings, this inequality remains true.

We estimate the Casson invariant v_2 of $\hat{\beta}$ using the Gauß diagram formula of Polyak-Viro (2.15). Let us put the basepoint right after the group of l positive crossings (which of either strands is immaterial).

Now l is odd and the l positive crossings are pairwise linked, so their contribution to the Gauß diagram sum (2.15) is, independently on the location of the basepoint, equal to $v_2(T_{2,l}) = \frac{l^2 - 1}{8}$, where $T_{2,l}$ is the $(2, l)$-torus knot. The analogous claim is true for the group of k negative crossings.

If some of the positive crossings is linked with a negative crossing, it must be a negative crossing of a σ_2^{-1}. Now by our choice of basepoint, in each syllable σ_2^{-2} only one of the two crossings gives a pair that contributes to (2.15), and the contribution is always $-1/2$. Putting this together, we have

$$v_2 \ge \frac{l^2 - 1}{8} + \frac{k^2 - 1}{8} - \frac{c_- - k + 2}{4} \cdot l \cdot \frac{1}{2}. \qquad (4.13)$$

Assuming $c \ge 10$ (the other cases are checked directly), we have

$$l \ge \frac{4c}{5} - 3 \ge \frac{c}{5} + 3 \ge c_- = c - l.$$

So the r.h.s. in (4.13) is minimized for $l \geq \frac{4c}{5} - 3$ when putting $l = \frac{4c}{5} - 3$. Then we have, with $k \geq 3$ and $c_- = c - l$, the estimate

$$v_2 \geq \frac{3c^2}{50} - \frac{29c}{40} + \frac{11}{4},$$

which is positive for $c \geq 10$, and grows when $c \to \infty$.

With this Theorem 4.3 is proved. $\square$

Remark 4.1 It is interesting whether max deg $Q \to \infty$ also for the strongly quasi-positive braids, but in Murakami's formula massive cancellations become possible and Q_1 cannot be easily controlled. At least one can prove using Theorem 4.5 and some results of Thistlethwaite in [Th] that there are only finitely many 3-braid links with given max deg$_z F$, where F is the Kauffman polynomial.

Chapter 5
Positivity of 3-Braid Links

5.1 Positive Braid Links

5.1.1 The Morton-Franks-Williams Bound

For the proof of Theorem 1.3 we will need to study the behavior of the bound (2.11) in the Morton-Franks-Williams inequality (we abbreviate as MFW) on positive braids. This was begun by Nakamura [Na], who settled the case $MFW = 2$ in the suggestive way: such braids represent only the $(2, n)$-torus links. (The case $MFW = 1$ is trivial.) We will introduce a method that considerably simplifies his proof (but still makes use of some of his ideas), and then go on to deal with $MFW = 3$. The example of non-sharp MFW inequality, 13_{9365} in [HT] (the connected 2-cable of the trefoil), given in [MS], is in fact only among a small family of exceptional cases.

Theorem 5.1 *If β is a positive braid and $MFW(\hat{\beta}) = 3$, then either*

1. $b(\hat{\beta}) = 3$ *and β reduces to a positive 3-braid, or*
2. $b(\hat{\beta}) = 4$ *and β reduces to a positive 4-braid given by one of the following forms (assuming that $3*$ denotes a sequence of at least one letter 3, and $11*$ resp. $22*$ sequences of at least two letters 1 or 2):*

$$[22*3*122*11*23211*], \quad and \quad [22*3122*11*23*211*]. \quad (5.1)$$

Definition 5.1 It might be useful to precisify already here what "reduce" should mean. We will simplify positive braid words (and the link diagrams corresponding under closure) using these types of (crossing number non-augmenting) transformations:

(1) the decomposition of diagrams (or words) as connected sum. If $\beta \in B_n$ can be written as $\beta = \alpha\gamma$ such that there is an $1 < i < n$ with α containing only σ_j

© The Author(s) 2017
A. Stoimenow, *Properties of Closed 3-Braids and Braid Representations of Links*,
SpringerBriefs in Mathematics, https://doi.org/10.1007/978-3-319-68149-8_5

for $j < i$, and γ only σ_j for $j \geq i$, then one can treat α and γ separately. This is meant to include the case $i = n - 1$ and $\gamma = \sigma_{n-1}$, when we have the deletion of nugatory crossings.

(2) the braid relations (commutativity and YB),
(3) the cyclic permutation of letters,
(4) and the *flype*. If $\beta = \alpha\sigma_1\gamma\sigma_1^k$, and σ_1 does not occur in α and γ, then β can be transformed into $\alpha\sigma_1^k\gamma\sigma_1$.

We will recur the proof to a finite number of words to check, which is done by calculation using the program of [MS]. Since a direct computation is more reliable than an increasingly difficult mathematical argument, we have not tried to minimize the calculation by all means. However, we point out that for the sake of Theorem 1.3 alone (rather than its refinement, Theorem 5.1), the following weaker statement is sufficient, for which a considerable part of the case-by-case calculations can be dropped. This corollary requires the notion of semiadequacy [LT], and can be deduced from Theorem 5.1 by direct check of the exceptional words. (We will sometimes write $MFW(\beta)$ for $MFW(\hat{\beta})$.)

Corollary 5.1 *If β is a positive braid word, and $MFW(\beta) = 3$, then β reduces (in the sense specified above) to a positive 4-braid word β', and the diagram $\hat{\beta}'$ is not B-adequate.*

Here the notions of A-adequate and B-adequate for diagrams and braids are as explained in Sections 2.2 and 2.5. We note, as a consequence of [Th], that a braid is $(A/B\text{-})$adequate if and only if some, or equivalently any, minimal length word of a braid in its conjugacy class it is so.

We will thus prove Theorem 1.3 only using Corollary 5.1, and indicate in the proof of Theorem 5.1 the point where the corollary follows (and the rest of the argument is not needed). The argument that elegantly replaces the remaining case-by-case checks requires Theorem 4.5. Since (by taking again the full extent of our proof) Theorem 4.5 is nonetheless not indispensable, we permit ourselves to defer its proof to the separate paper [St4].

Proof of Theorem 1.3 If $MFW(\beta) \leq 2$, then we are done. So assume $MFW(\beta) = 3$. Assume first L has an A-adequate 3-braid. Since A-adequate diagrams minimize the number of negative crossings, and L has a positive (braid) diagram, the A-adequate 3-braid diagram is positive, and we are done. So let L have a B-adequate 3-braid diagram $\hat{\beta}'$. Now by Corollary 5.1, we find that L reduces to a positive 4-braid β with $c_+(\beta)$ positive crossings, and β is not B-adequate. Since β' is B-adequate and B-adequate diagrams minimize the number of positive crossings, $c_+(\beta') < c_+(\beta)$. On the other hand, $MFW(L) = 3$ and the inequalities of MFW for the v-degree of P show that β' must have exponent sum $[\beta'] = c_+(\beta) - 1$. Since $[\beta'] \leq c_+(\beta')$, we must have equality, so β' is positive.   □

Note the following easy and useful consequence of Theorem 1.3:

Corollary 5.2 *A link which is a closure of a positive braid of at most four strings has a minimal crossing diagram as a closed positive braid, and a minimal string positive braid representation.*

Proof The case of the braid representation is straightforward, and it implies the minimal crossing diagram statement by looking at $\max \deg_z P$ and using Morton's inequalities. □

The examples in [St3], mentioned after Theorem 1.3, show that the corollary is not true in case of positive 5-braids at least for the positive minimal braid representation. So far no examples are known where no minimal crossing positive braid diagram exists (it was known to exist from [FW, Mu] for closed positive braids with a full twist, which include the torus links, and from [St3] for positive braid knots of at most 16 crossings), but the pathologies for minimal strings hint to caution. It was shown in [St3] that one, and in [St4] that infinitely many fibered positive knots have no minimal crossing positive diagram.

5.1.2 Maximal Subwords

Here we start the technical considerations needed to prove Theorem 5.1. We consider the form (2.9), now with all $l_i > 0$.

Definition 5.2 We define *summit syllables* in (2.9):

(a) All $\sigma_{n-1}^{l_i}$ are summit syllables, and
(b) if $\alpha = \sigma_k^{l_i}$ and $\alpha' = \sigma_k^{l_j}$ are summit syllables, with no $\sigma_{k'}^{l'}$ for $k' \geq k$ occurring between α and α', then all $\sigma_{k-1}^{l'}$ occurring between α and α' are summit syllables.

Note that, according to Definition 2.1, we consider syllables in cyclic order. The relation "between" in the above definition should also be understood in that sense: a syllable occurring after the last index i syllable α and/or before the first index i syllable α' is considered to be between α and α'.

For the following considerations it is (not necessary but) helpful to visualize β by the (minorly modified) braid scheme explained in [St6].

Separate β in (2.9) into subwords $\alpha_1 \ldots \alpha_n$, such that α_i contains only syllables of odd or even index, and this parity changes between α_i and α_{i+1}. Then for a syllable σ_k^l occurring in α_i, put the integer l at the point $(k, -i) \in \mathbb{N} \times \mathbb{Z} \subset \mathbb{R}^2$ in the plane. Here $(k, -i)$ is the point in the ith row and kth column, with rows numbered (as in Cartesian coordinates) from bottom to top and columns from left to right.

One obtains a certain checkerboard pattern of integers we call *braid scheme* of β. (If we do not put any integer on a point $(k, -i)$, we assume its "content" is zero, or it is "empty." So for all non-empty points $(k, -i)$ in the scheme, $i + k$ is always even or always odd.)

In a scheme, summit syllables of β are those, whose entries in the scheme are "on top" when viewing the scheme from the left. They are shown slanted in the below example:

$$[1234132112324321] \quad = \quad \begin{array}{ccccccc} & & 4 & & & & 4 \\ & 3 & & 3 & & 3 & & 3 \\ 2 & & & 2 & & 2 & & 2 & & 2 \\ 1 & & & 1 & & 1\ 1 & & & 1 \end{array}.$$

$$\tag{5.2}$$

From this viewangle the following "geographic" choice of terminology becomes more plausible.

Definition 5.3 Summit syllables still have a cyclic order from (2.9). We call the subword β' of β in (2.9) made of summit syllables the *maximal subword*. The subword made of non-summit syllables (i.e. the subword obtained by deleting in β all syllables in β') is called *non-maximal subword*.

Note that neighbored summit syllables have indices k_i differing by ± 1. We say that a summit syllable is *minimal* resp. *maximal* if its both neighbors have higher resp. lower index.

We call β *summit reduced* if all its minimal summit syllables are non-trivial. We call β *index reduced* if it is non-singular and its index sum $\sum_{i=1}^{n} k_i \cdot l_i$ cannot be reduced by a Yang-Baxter relation, i.e. β contains no $\sigma_{i+1}\sigma_i\sigma_{i+1}$ as subword.

Lemma 5.1 *Index reduced $\Longrightarrow$ summit reduced. In particular a summit reduced form always exists.* □

Recall that a positive resolution tree is a rooted tree with directed edges, whose vertices (nodes) contain positive braid words, the root labelled by β. Every vertex has exactly one incoming edge, except the root that has none, and zero or two outgoing edges. In former case it is labelled by an unlink (terminal node). In latter case it is labelled by a word of the form $\alpha\sigma_i^2\alpha'$, with α, α' positive words, and the two vertices connected by the outgoing edges are labelled by $\alpha\sigma_i\alpha'$ and $\alpha\alpha'$, or positive words obtained therefrom by Markov equivalence (isotopy of the closure link).

In [Na] the following fact was observed, and used decisively, and we shall do the same here.

Theorem 5.2 (Nakamura [Na]) $MFW(\beta)$ *is the maximal number of components of a (link in a) node in a positive resolution tree for β.*

In particular, MFW is monotonous (does not decrease) under word extension, and does not depend on the exponent of non-trivial syllables.

Lemma 5.2 *If β is summit reduced, and β' is obtained from β by removing all summit syllables, then there is a positive resolution tree for β that contains β' as a node.*

Proof Since minimal syllables are non-trivial, one can delete them in the resolution tree. The two neighbors in the maximal subword join to a non-trivial new minimal syllable, and so one iterates the procedure.   □

Now we have a quick proof of Nakamura's main result.

Corollary 5.3 (Nakamura [Na]) *If a positive braid β has $MFW(\beta) = 2$, then $\hat{\beta}$ is a $(2, n)$-torus link.*

Proof Since all $\sigma_{n-1}^{\nu'}$, $\sigma_{n-2}^{\nu'}$ in (2.9) occur as summit syllables, β' has split last two strands. Therefore, any summit reduced positive word on $n \geq 3$ strands has $MFW \geq 3$.   □

In particular any positive braid representation of a $(2, n)$-torus link can be reduced to the standard one by the first three operations given in Definition 5.1 (i.e., the flype is not needed), with the first operation restricting to the removal of nugatory crossings.

5.1.3 The Proof of Theorem 5.1: Initial Simplifications

The following fact is well-known:

Theorem 5.3 $MFW(\beta) = 1$ *if and only if* $[\beta]_i = 1$ *for* $i = 1, \ldots, n - 1$.

Now for $MFW = 3$ it suffices to ensure that (either we can reduce the braids) or can find words, whose non-maximal subwords do not give an unknot (excluding the two rightmost isolated strands). *For the rest of the chapter we assume that β' gives the unknot.*

We will work by induction on the number of strands, and for fixed number of strands on the index sum. So we consider a positive braid word β, and assume w.l.o.g. it has the smallest index sum among positive braid representatives of its closure link for the same number of strands. For such β, we will either reduce it (by at least one strand or crossing) or show $MFW \geq 4$.

Most braids β will be easily dealt with, but there remain certain families of words that require a case-by-case study. We decided not to omit too many of the (tedious) details of this part, in order to keep the proof followable, even if it may not contribute to its (esthetic) appearance.

Note that in order to prove $MFW(\beta) \geq 4$, it suffices to go over to a (link in a) suitably chosen node in a positive resolution tree for β and show $MFW \geq 4$ for this node. In particular, we can remove from β all syllables of index $\leq k$ and the k resulting left isolated strands.

The case of reductions is more delicate, but with some care we will be able to describe them directly.

The following is a good tool for showing $MFW \geq 4$. It, and some of the subsequent appeal to geography, is again motivated by the schemes as in (5.2), which we will occasionally use.

Definition 5.4 A *valley* resp. *mountain* is a subword of the maximal word that starts with same index syllables and contains only one minimal resp. maximal syllable. This syllable is called the *bottom* of the valley resp. *summit* or *top* of the mountain. The index of the bottom/top is the *depth* resp. *height*.

We assume there are at least two mountains of maximal height (i.e., $n - 1$). Otherwise we have a split component or a $(2, n)$-torus connected component, or a reducible braid. All of these can be dealt with by induction on the number of strands. Similarly at least one valley has depth 1, otherwise σ_1 in β' remains reducible in β.

The following operation will be somewhat important, and we will call it "filling the valley."

Lemma 5.3 ("Filling the Valley") *Any valley can be removed from the maximal subword in the positive resolution tree.*

Proof Same as for Lemma 5.2.   □

Lemma 5.4 *If a mountain M is not of maximal height (i.e., $n - 1$), then $MFW(\beta) \geq 4$.*

Proof Let $k < n-1$ be the height of M. We assumed there are at least two mountains of maximal height. So now w.l.o.g. assume some, say the left, of the neighbored mountains of M has height $k' > k$. Fill the two valleys around M starting with σ_{k-1}. Then the maximal subword has a syllable index sequence $k + 1, k, k - 1, k, k - 1, k$. Make the second and third syllable trivial (if not already), and apply YB relations, moving the fourth syllable to the left: $k + 1, \underline{k - 1}, k, k - 1(, k - 1), k$. The result is a summit reduced word, in which a new syllable of index $k - 1$ (the underlined one) was removed and it became non-maximal. Hence the non-maximal subword has exponent sum $[\beta']_{k-1} > 1$, and so $MFW \geq 4$.   □

Lemma 5.5 *If β has > 2 valleys of depth at most $n - 3$, then $MFW(\beta) \geq 4$.*

Proof It suffices to check for three mountains (as one can fill separate valleys) and 4-braids (as one can fill valleys by levels as in the proof of Lemma 5.2 and the remark after Theorem 5.3). This is just the word [1232112321121321], which is easily checked (to have $MFW = 4$).   □

Definition 5.5 Let us here fix the *complexity* with respect to which we will simplify braid words.

- First, braids of fewer strands are simpler.
- Among (braids of) equal number n of strands, we count the number of syllables of index $n-1$ (which by Lemma 5.4 may be assumed to be equivalent to mountains). The words with fewer such syllables (or mountains) are simpler.
- Finally, for equal number of strands and mountains, we consider the summit reduced words as simpler than those which are not.

5.1.4 Two Mountains

We assume in Sections 5.1.4 and 5.1.5 that $n \geq 5$. The case of 4-braids is considered later in Section 5.1.6. We refer to Section 2.4 for the use of notation we will employ.

We assume first β has two mountains. By Lemma 5.4 they may be assumed to be both of height $n - 1$.

So now consider words with syllable index sequence

$$1, 2, \ldots, n-2, n-1, \quad p_1@, \ldots, p_k@, \quad n-2, n-3, \ldots, g+1, g!,$$

$$g+1, \ldots, n-2, n-1, \quad q_1@, \ldots, q_l@, \quad n-2, n-3, \ldots, 2, 1 \qquad (5.3)$$

such that $l + k = n - 3$ and $\{p_1, \ldots, p_k, q_1, \ldots, q_l\} = \{1, \ldots, n-3\}$. We will distinguish only between non-trivial and trivial syllables (in former case exponent is immaterial). For non-trivial syllables we write an exclamation mark after the index, for trivial ones an "at" (@) sign. If none of ! and @ is specified, we do not exclude explicitly any of either types. We write $\beta_1, \ldots, \beta_6$ for the subwords separated by space in (5.3).

Assume w.l.o.g. (up to reversing the braid's orientation) that some p_i is 1, and let up to commutativity the $p_1, \ldots, p_k$ subword be written as $h, h-1, \ldots, 1, p'_1, \ldots, p'_{k-h}$ (the p'_i contain the indices above h occurring as p_i).

For the case distinction below, we will prepare the following argument. Assume the maximal syllable of σ_{h+1} in β_1 is non-trivial. Then one can write β as word with a subword of index sequence

$$1, \ldots, h+1!, h@, \ldots, 1@ . \qquad (5.4)$$

If now $h < n - 3$, we can make in (5.4) all syllables trivial except $h + 1$, which we make of exponent 2, then split a loop (component of a link in a node of a positive resolution tree) by removing all the syllables in (5.4) together with the terminating "1" in (5.3). We obtain

$$h+2, \ldots, n-2, n-1, \quad p'_1@, \ldots, p'_{k-h}@, \quad n-2, n-3, \ldots, g+1, g!,$$

$$g+1, \ldots, n-2, n-1, \quad q_1@, \ldots, q_l@, \quad n-2, n-3, \ldots, 2.$$

(Here no syllables of index 1 occur, and the split loop is the isolated leftmost strand.) Then we fill the valley starting with the index-$n-1$-summits, splitting another loop (the rightmost strand),

$$h+2, \ldots, n-3, n-2, \quad p'_1@, \ldots, p'_{k-h}@, \quad q_1@, \ldots, q_l@, \quad n-2, n-3, \ldots, 2,$$

and are left with a word that has at least two σ_{n-2}. So $MFW \geq 4$.

Case 1. Assume the smaller valley has depth $g > 1$. The word β has two index-1 syllables, a trivial (non-summit) and a non-trivial (summit) one. By a flype one can exchange them, and so have a non-summit reduced word. Then one can change β to a word of smaller index sum, and so we are done by induction.

Case 2. Now $g = 1$. We write $\beta_1, \ldots, \beta_6$ for the six subwords separated by spacing in (5.3). By a similar argument as after (5.4) we can argue that if one can reorder the syllables in $\beta_{2,5}$ so that β has a subword with an index sequence

$$k, k+1, \ldots, h-1, h!, h-1, \ldots, k, \qquad (5.5)$$

with the first or last $h - k$ syllables belonging to $\beta_{2,5}$ and the others to $\beta_{1,3,4,6}$, and $k < h < n - 2$, then $MFW \geq 4$. (We will make use of this argument several times below.)

W.l.o.g. assume $1 \in \beta_2$ (which is meant to abbreviate that β_2 contains an index-1 syllable).

The form (5.3) implies that there are ≤ 3 syllables of index 1 in β. One can commute the non-maximal syllables so that

$$1!\, 2\, 1@ \quad \text{or} \quad 1@\, 2\, 1! \qquad (5.6)$$

is a subword of β, with the "$1@$" being the non-maximal syllable of β_2, and the other two syllables being maximal. With the notation (5.2), let us follow the typical example

$$
\begin{array}{cccccccc}
 & & \mathbf{4} & & & & \mathbf{4} & \\
 & \mathbf{3} & & \mathbf{3} & & \mathbf{3} & & \mathbf{3} \\
 & \mathbf{2} & & & \hat{2} & \mathbf{2} & & \mathbf{2} & & \mathbf{2} \\
 1 & & & 1 & & 1\ 1 & & & 1
\end{array}
\qquad (5.7)
$$

Case 2.1. If the index-2 syllable in (5.6) (which is distinguished by a hat in (5.7)) is non-trivial, we have (5.5), and $MFW \geq 4$ with the argument thereafter.

Case 2.2. The exclusion of (5.5) means that

$$1!\, 2@\, 1@ \quad \text{or} \quad 1@\, 2@\, 1! \qquad (5.8)$$

is a subword of β, with the "$1@$" being the non-maximal syllable.

Case 2.2.1. Assume both (bold) index-4-syllables in (5.7) are non-trivial. Then one can delete first the index-1 syllables, then delete the index-4 syllables, splitting one loop, and has an extension of [232323], with $MFW = 3$. Thus in total $MFW \geq 4$. (For general n replace 4 by $n - 1$, 3 by $n - 2$, 2 by $n - 3$, and delete all syllables of index 1 through $n - 4$ inclusively.)

Case 2.2.2. We assume thus now some index-4-syllable in (5.7) is trivial.

We are going now to describe a fundamental *pattern of reductions* of the complexity of braid words (in the sense of Definition 5.5), which will occur throughout the rest of the proof.

We now apply YB relations to (5.8), making the "1!" to an index-2 syllable. This temporarily spoils the summit-reducedness, but we will restore it in another way. The YB relations have left only two index-1 syllables, one of which is trivial, and we can exchange them by a flype.

This makes out of (5.7) the following picture:

$$
\begin{array}{cccccccc}
 & \mathbf{4} & & & & & \mathbf{4} & \\
 & \mathbf{3}\ \ \mathbf{3} & & & & \mathbf{3}\ \ \mathbf{3} & & \\
\mathbf{2} & & \underline{2}\,\underline{2} & & \underline{2}\,\underline{2} & & \mathbf{2}\ \ \mathbf{2} & \\
 & 1 & & 1\ 1 & & & &
\end{array}
\tag{5.9}
$$

One should notice that the application of YB relations on (5.6) means that the (underlined) index-2 syllables neighbored to the non-trivial index-1 syllable are both non-trivial. Notice also that the alternative

$$
\begin{array}{cccccc}
 & \mathbf{4} & & & \mathbf{4} & \\
 & \mathbf{3}\ \ \mathbf{3} & & & \mathbf{3}\ \ \mathbf{3} & \\
\mathbf{2} & & \mathbf{2}\ \ \hat{\mathbf{2}} & & \mathbf{2} & \mathbf{2} \\
 & 1 & & 1\ \ 1\ 1 & & 1
\end{array}
$$

to (5.7) would lead to the same picture (5.9).

Case 2.2.2.1. If now the two bold index-2 syllables in (5.9) are both non-trivial, then (by the remark after (5.9)) all 4 maximal index-2 syllables are non-trivial. We could then delete them, and obtain a summit-reduced word with more than one non-maximal letter "1". Thus $MFW \geq 4$.

Also, if the two bold index-3 syllables are both non-trivial, we could delete these two syllables, again obtaining a summit-reduced word with more than one non-maximal letter "1". (For general n, this argument applies for index $3 \leq n' \leq n - 2$.)

Case 2.2.2.2. We have thus now that for each $2 \leq n' \leq 4$ (or $2 \leq n' \leq n - 1$ for general n), one of the two bold-faced syllables of index n' is trivial.

Now we can get a reduction of the number of syllables of index 4 (or $n - 1$). To show it, we display it in one example, and only a part of the word of (5.9) after cyclic permutation. (In the leftmost upper picture below, the syllables of index $2, 3, 4$ are those set in bold in (5.9).)

$$
\begin{array}{ccc}
\begin{array}{cccc}
4 & & & \\
 3\ 3 & & & 4\ 4 \\
 & 2\ 2\ 2 & & 3 \\
 & 1 & &
\end{array}
&
\longrightarrow
&
\begin{array}{ccc}
4 & & 4\ 4 \\
 3\ 3 & 3 & \\
 & 2 & \\
1\ 1 & 1 &
\end{array}
\end{array}
\longrightarrow
$$

$$
\longrightarrow
\begin{array}{ccc}
4\ \ 4\ 4 & & \\
 3 & & \\
 2\ \ 2\ 2 & & \\
1\ 1 & & 1
\end{array}
\longrightarrow
\begin{array}{ccc}
 & 4 & \\
 & 3\ 3\ \ 3 & \\
1\ 1 & 2 & 2\ 2 \\
 & & 1\ 1
\end{array}
\ \ \cdot
\tag{5.10}
$$

This finishes the case of two mountains.

5.1.5   More than Two Mountains

To deal with the general case, now we make the following modifications. We call a summit syllable sequence with indices $n-2$ and $n-1$ terminated on both sides by $n-1$'s a modified mountain or *plateau*. We have again by Lemma 5.5 only two valleys of depth $< n-2$, or alternatively only two plateaus (now instead of mountains). The case of more than two mountains is thus mainly an adaptation of the case of two mountains, replacing mountains by plateaus.

Again we may assume non-maximal subwords have exactly one (and trivial) syllable per index. The elimination of the maximal subwords can be done similarly.

We distinguish two cases as in the above study of the 2-mountain words, depending on the depth g of the second valley (the other valley has depth 1 by the same argument as above).

Case 1. $g > 1$. We use the previous flyping argument.

Case 2. $g = 1$. In the second case we have again (5.8), because $n-2 \geq 3$, and we can commute a non-maximal index-1 syllable to be neighbored to a valley, as in (5.7). The rest of the argument continues in a similar fashion.

This completes the proof for $n \geq 5$ strands.

5.1.6   4-Braids

We are now down to the considerably more explicit case of four strands.

If some mountain is not of height 3, or > 2 valleys of depth 1 exist, then we are done as before (see Section 5.1.3). If we have two valleys, with one of them of depth > 1, we can use again the argument in Case 1 of Section 5.1.4.

So the maximal subword is of the form[1] $1, 2, 3, (2!, 3)^p, 2, 1, 1, 2, 3, (2!, 3)^n, 2, 1$, and the non-maximal subword is a single σ_1^1. We separate the summit syllables and their letters by the summit syllables of index 1 into a *left* and *right plateau*. Assume w.l.o.g. the (non-summit) σ_1^1 is in the left plateau. The word "in" is to mean that in cyclic order of the syllables of β the syllable σ_1^1 can be written to occur just before or after a syllable with index 3 that belongs to the left plateau. This means that we can write β as

$$\beta_{l,m,n} = [123(223)^n 1(223)^m 21123(223)^l 21] \tag{5.11}$$

with $n, m, l \geq 0$, or some of its extensions. Note that MFW will be monotonous in m, n, l, i.e. $MFW(\beta_{l,m,n+1}) \geq MFW(\beta_{l,m,n})$, etc. Using symmetry assume $n \geq m$.

---

[1]Notice we use n in a different meaning from before!

One can check already at this stage that such words are not B-adequate. So we obtain Corollary 5.1, and for the proof of Theorem 1.3 the rest of the argument here can be replaced by the application of Theorem 4.5. Note that B-adequacy is invariant under isotopy preserving writhe and crossing number, so any other positive 4-braid word giving the same link is not B-adequate either.

If $n + l > 0$, then the 2-index syllable 2 in (5.11) must be trivial. Otherwise remove all 2! in $(2!, 3)^m$ (if any), and split a loop from the first strand, as explained after (5.4). You remain with a word in 2 and 3, which is an extension of [232323] of $MFW = 3$. In total thus $MFW = 4$. With a similar argument we see that if $m + l > 0$, then the 2-index syllable 2 is trivial.

We distinguish three cases depending on whether these arguments apply or not, i.e. whether $n + l > 0$ and/or $m + l > 0$.

Case 1. Both non-triviality arguments apply. So we have a family of words $\beta_{l,m,n} = [123(223)^n 1(223)^m 21123(223)^l 21]$ with $n, m > 0$ or $l > 0$, and their extensions, and both 2 and 2 are trivial.

Case 1.1. $l = 0$. We assumed $n, m > 0$, and already for $n = m = 1$, the word [12322312223112321], we have $MFW = 4$.

Case 1.2. $l > 0$.

Case 1.2.1. $m = n = 0$. These are extensions of $[1231211123(223)^l 21]$.

The two letters "2" in the left plateau cannot be doubled ([123122112322321], [122312112322321]), neither the "3" ([123132112322321]), since $MFW = 4$ already for $l = 1$.

In the remaining cases, we find that $[12312112 * 3[23] + 2 * 1]$ reduce w.r.t. the number of syllables of index 3, using the simplification described in Case 2.2.2 of Section 5.1.4. (Recall that, while "2*" in a braid word should mean at least one letter "2", the term "[23]+" should mean a possibly empty sequence of letters "2" and "3". We distinguish braid words from index sequences by not putting commas between the numbers.)

Case 1.2.2. $m + n > 0$; this reduces to the case of $m = n = 0$ with some of the twos or the three in the left plateau doubled, where we found $MFW = 4$.

Case 2. In the case one of the non-triviality conditions on 2 and 2 does not apply, we have $[123(223)^n 12112321]$ with $n > 0$ and its extensions. (Now the right plateau is a mountain; the case $m > 0, n = 0$ is symmetric.)

Since $MFW([1232231211123221]) = 4$, the right 2-index syllable in the left plateau (2 in (5.11)) must be trivial.

Also, the right "2" and the "3" in the right plateau must be trivial: [1232231211123221] and [1232231211123321] have $MFW = 4$. (But the left "2" of the left and right plateau may not be trivial.)

Then $[12 * [23] + 312112 * 321]$ simplifies as in Case 2.2.2 of Section 5.1.4.

Case 3. In case neither non-triviality condition applies, i.e. $n = m = l = 0$, we have [12312112321] and its extensions. (So both plateaus are mountains.)

Since non-triviality is nonetheless possible, we may have non-trivial 2-index syllables in the left plateau. We distinguish three cases again according to whether 2 and 2 in (5.11) are trivial or not.

Case 3.1. Both 2-index syllables 2 and 2 are trivial: $[1\hat{2}31\hat{2}112321]$. (A hat is to mean that the syllable may not be extended.)

By direct check (or by arguments we had) $[1233121123321]$ has $MFW = 4$. So one can extend the right mountain's "2"s and one of the left or right mountain's "3"s, but not both "3"s.

A reducing check is therefore to be made on $[11 * 231211 * 22 * 3 * 22*]$ and $[11 * 23 * 1211 * 22 * 322*]$. Both can be handled using the aforedescribed simplification.

Case 3.2. One 2-index syllable is non-trivial. This is the word $[1\hat{2}3122112321]$ and its extensions (the case $[122312\hat{2}112321]$ is symmetric).

Among the extensions (except for the hatted "2"), $[1231221122321]$ has $MFW = 4$, as well as $[12331221123321]$.

Thus we are left to deal with $[1233 * 122 * 11 * 2322 * 1*]$ and $[123122 * 11 * 233 * 22 * 1*]$, and check that they both reduce as before.

Case 3.3. Both 2-index syllables are non-trivial: $[1223122112321]$. This is a braid word β_0 for 13_{9465}. We have the following extensions:

$[12233122112321]$ has $MFW = 3$,

$[12231221122321]$ has $MFW = 4$,

$[12231221123221]$ has $MFW = 4$,

$[12231221123321]$ has $MFW = 3$.

The only common extension of the two $MFW = 3$ extensions is $[122331221123321]$, which has $MFW = 4$.

The words that remain to treat are $[22 * 3 * 122 * 11 * 23211*]$ and $[22 * 3122 * 11 * 23 * 211*]$, given in (5.1). So it is to verify that their closures have braid index 4. For this we use the two-cabled MFW inequality. Let for a braid $\beta \in B_n$, the "two-cabled" braid $(\beta)_2 \in B_{2n}$ be obtained from β by replacing in (2.9) each σ_i by $\sigma_{2i}\sigma_{2i-1}\sigma_{2i+1}\sigma_{2i}$. Then $(\beta)_2$ is a braid representation of the (blackboard framed) two-cable link L_2 of the closure $L = \hat{\beta}$ of β. We consider now for β the above braid β_0.

We know, from the computations described in [MS, FW], that $MFW((13_{9465})_2) = 7$. This is in fact true also for the connected cable (the one with braid representation $(\beta_0)_2 \cdot \sigma_1$). Now we claim that reducing or resolving a clasp (changing a σ_i^2 into a σ_i or deleting it) does not reduce the two-cabled MFW bound. The 2-cable of a clasp can be resolved by resolving four clasps. The 2-cable of a crossing in a 2-cabled clasp can be resolved by resolving one clasp and changing twice $\sigma_i^2 \to \sigma_i$. Finally, the 2-cable of an isolated σ_i can be reduced into two internal twists of the doubled original strand. Such twists can be collected for every doubled component, resolved for each doubled component to one, and joined if doubled components are joined by reducing a doubled crossing in a doubled clasp. So the two-cabled MFW reduces to the one of the connected cable of 13_{9465} and we are done.

The proof of Theorem 5.1 is now completed.

5.2 Positive Links

In this section, we will refine the arguments proving Theorem 1.1 to restrict the possible 3-braid representations of positive links. Our positivity considerations will make use of the criterion of Yokota [Yo], and the Kauffman polynomial F. We recall the properties (2.6)–(2.8) that determine F and its writhe-unnormalized version Λ.

Note that for P one can similarly define a regular isotopy invariant

$$\tilde{P}(D)(a, z) = (ia)^{-w(D)} P(D)(ia, iz), \qquad (5.12)$$

with $i = \sqrt{-1}$. Then $\tilde{P}$ satisfies similar relations to (2.6)–(2.8). The difference to Λ is that $\tilde{P}$ is defined on oriented link diagrams, and that the term making orientation incompatible on the right of (2.6) is missing.

Theorem 5.4 (Yokota [Yo]) *If L is a positive link, then*

$$\min \deg_a F(L) = \min \deg_v P(L) = 1 - \chi(L),$$

and

$$[F(L)]_{a^{1-\chi(L)}} = [P(L)(ia, iz)]_{a^{1-\chi(L)}}.$$

We also require an extension of braids to the context of F. This was described in [BW] and [Mr], but we use only the generators of the algebra defined there. Strings will be assumed numbered from left to right and words will be composed from bottom to top. We write σ_i for a braid generator, where strand i from the lower left corner, passing over strand $i + 1$, goes to the upper right corner.

We add elements δ_i of the following form:

By hat we denote the usual closure operation.

A word in the described generators gives rise to an unoriented tangle diagram that turns into an unoriented link diagram under closure. If the word has no δ_i then this diagram can be oriented to give a(n oriented) closed braid diagram. We will assume this orientation is chosen. Otherwise, a coherent orientation is not generally possible. In particular, the sign of exponents of σ_i in this context may not coincide with the sign of the corresponding crossings after some (or even any) orientation choice of the diagram. For kinks (the diagram fragments occurring in (2.7) on the left hand-sides), however, a sign is definable since any possible component orientation chosen gives rise to the same (skein) sign. So we will be able (and we will need) to distinguish between positive (in the left equation of (2.7)) and negative kinks (in the right one).

Lemma 5.6 *Let*

$$D = \widehat{} \left(\sigma_1^{k_{1,1}} \sigma_2^{k_{1,2}} \dots \sigma_1^{k_{1,n_1}} \sigma_2^{-1} \quad \sigma_1^{k_{2,1}} \dots \sigma_1^{k_{2,n_2}} \sigma_2^{-1} \quad \dots \quad \sigma_l^{k_{l,1}} \dots \sigma_1^{k_{l,n_l}} \delta_2 \right),$$
(5.13)

where $n_i \geq 1$ odd and $k_{i,j} \geq 2$ when $1 < j < n_i$ and $k_{i,j} \geq 1$ when $j = 1$ or n_i. Then
$\min \deg_a \Lambda(D) = -l$.

Proof For $l = 1$ see Lemma 5.9 below. Then use induction on l. Change a crossing of a σ_2^{-1} in $D = D_-$. Then:

- D_+ can be reduced by the same argument as in the proof of Lemma 6.2 of [St4] until we have a form (5.13) with smaller l, and a non-zero number of negative kinks added. Then since negative kinks shift the a-degree of Λ up (according to the second formula in (2.7)), we have $\min \deg_a \Lambda(D_+) > -l$.
- D_0 has $\min \deg_a \Lambda(D_0) = 1 - l > -l$ by induction.
- $D_\infty = D_{l_1} \# D_{l_2}$ with $l_1 + l_2 = l$, so $\min \deg_a \Lambda(D_\infty) = -l_1 - l_2 = -l$, and $\min \deg_a \Lambda(D)$ is inherited from $\Lambda(D_\infty)$. $\square$

Lemma 5.7 *Let*

$$D = \widehat{} \left(\sigma_1^{k_{1,1}} \sigma_2^{k_{1,2}} \dots \sigma_1^{k_{1,n_1}} \sigma_2^{-1} \quad \sigma_1^{k_{2,1}} \dots \sigma_1^{k_{2,n_2}} \sigma_2^{-1} \quad \dots \quad \sigma_l^{k_{l,1}} \dots \sigma_1^{k_{l,n_l}} \sigma_2^{-1} \right),$$
(5.14)

where $n_i \geq 1$ odd and $k_{i,j} \geq 2$ when $1 < j < n_i$ and $k_{i,j} \geq 1$ when $j = 1$ or n_i. Then
$\min \deg_a \Lambda(D) \geq -2$ *when $l \leq 2$ and* $\min \deg_a \Lambda(D) = -l$ *when $l \geq 3$.*

Proof Assume first we proved the result for $l \leq 2$, and that $l > 2$. We argue by induction on l.

Apply the Λ-relation at a σ_2^{-1} crossing in $D = D_-$. Then D_0 has $\min \deg_a \Lambda = 1 - l$ by induction. D_+ simplifies as in the proof of Lemma 6.2 of [St4]. This simplification only removes negative letters. It can be iterated until one of two situations occurs. It can (a) happen that all negative letters disappear. Then we have a positive braid and $\min \deg_a \Lambda = -2 > -l$ by Yokota's result. Or it can (b) occur that no $\sigma_1\sigma_2\sigma_1$ or $\sigma_2\sigma_1\sigma_2$ occur as subwords. Then $k_{i,j} \geq 2$ for $1 < j < n_i$, and the number l of negative crossings has decreased strictly. So we have by induction $\min \deg_a \Lambda(D_+) > -l$. Finally we must deal with the D_∞ term. This follows from Lemma 5.6.

It remains to justify the claim $\min \deg_a \Lambda(D) \geq -2$ when $l \leq 2$. This is done exactly with the same argument, only that now we observe that all of $\Lambda(D_{0,+,\infty})$ have a-degree ≥ -2. $\square$

Lemma 5.8 *The closed braids in* (5.14) *are not positive links for $l \geq 2$ and $n_i \geq 3$.*

Proof The representation (5.14) clearly gives rise to a positive band representation by replacing $\sigma_2\sigma_1^{k_{i,n_i}}\sigma_2^{-1}$ by $\sigma_3^{k_{i,n_i}}$. So $[\beta] = 3 - \chi(\hat{\beta})$. If $l > 2$, then we have

$$\min \deg_a F(D) = w(D) + \min \deg_a \Lambda(D) = 3 - \chi(\hat{\beta}) + \min \deg_a \Lambda(D) < 1 - \chi(\hat{\beta})$$

but min $\deg_v P(D) = 1 - \chi(\hat{\beta})$ as before, so min $\deg_a F \neq$ min $\deg_v P$ and we are done by Theorem 5.4.

To deal with $l = 2$, we consider first $l = 1$. Then if $n_1 = 1$, we have a $(2, k_{1,1})$-torus link, and for $n_1 = 3$ we have the $(1, k_{1,1}, k_{1,2}, k_{1,3})$-pretzel link. Otherwise we apply the relations for $\tilde{P}$ and Λ at the negative crossing. Then min $\deg_a \Lambda(D_{+,0}) =$ min $\deg_a \tilde{P}(D_{+,0}) = -2$, and $[\Lambda(D_{+,0})]_{a^{-2}} = [\tilde{P}(D_{+,0})]_{a^{-2}}$ by Theorem 5.4 since $D_{+,0}$ are positive braids. That the extra term $\Lambda(D_\infty)$ has no contribution to a^{-2} follows from Lemma 5.9 below. So D satisfies Yokota's conditions in Theorem 5.4. (We do not always know if D depicts a positive link; see Remark 5.2.)

If $l = 2$, then resolve a negative crossing in $D = D_-$ via the relation $\tilde{P}_- = z\tilde{P}_0 - \tilde{P}_+$ and via the Λ-relation $\Lambda_- = z\Lambda_0 + z\Lambda_\infty - \Lambda_+$. Now $[\tilde{P}_{+,0}]_{a^{-2}} = [\Lambda_{+,0}]_{a^{-2}}$. This follows from the above argument for $l = 1$. The additional term Λ_∞ has a-degree -2 by Lemma 5.6, so $[\Lambda_-]_{a^{-2}} \neq [\tilde{P}_-]_{a^{-2}}$. Again by Theorem 5.4, $D = D_-$ can therefore not belong to a positive link. $\qquad\square$

Remark 5.1 Ishikawa asks in [I] whether strongly quasi-positive knots satisfy the equality $TB = 2g_s - 1$, where TB is the maximal Thurston-Bennequin invariant. The proof of the lemma shows that there are infinitely many strongly quasi-positive 3-braid knots with $TB < 2g_s - 1$: take any of the knots with $l > 2$. (There are also many other such knots, like the counterexamples to Morton's conjecture in [St3].)

Lemma 5.9 *If*

$$D = D_{[n]} = \frown(\sigma_1^{k_1}\sigma_2^{k_2}\ldots\sigma_1^{k_n}\delta_2)$$

with $n \geq 1$ odd, $k_1, k_n \geq 1$ and $k_i \geq 2$ when $i = 2, \ldots, n-1$, then min $\deg_a \Lambda(D) = -1$. *Writing $\bar{k} = (k_1, \ldots, k_n)$ and*

$$\tilde{w}(\bar{k}) = (k_1 - 1) + \sum_{l=2}^{n}(k_l - 2),$$

we have max $\deg_z[\Lambda(D)]_{a^{-1}} = \tilde{w}(\bar{k})$.

Proof If $n = 1$ we check directly (we have a reduced diagram of the $(2, k_1)$-torus link), so let $n \geq 3$.

Consider first three special forms of $\bar{k}$.

If $\bar{k} = (12^*1)$ (with 2^* being a sequence of "2"), then D is regularly isotopic to a trivial 2-component link diagram. If $\bar{k} = (12^*)$ or (2^*1), then D is regularly isotopic to an unknot diagram with one positive kink. In both situations the claims follow directly.

Now let $\bar{k} = (2^*)$. We orient D so as to become negative. Then D depicts the $(2, -n - 1)$-torus link. We can evaluate Λ on L from $F(L)$ by normalization. For its mirror image $!L$, we can use Theorem 5.4 to conclude that min $\deg_a F(!L) = n$. Now it is also known that $\text{span}_a F(!L) = c(!L)$, and $c(!L) = n + 1$, so

$$\text{max} \deg_a F(!L) = \text{min} \deg_a F(!L) + \text{span}_a F(!L) = n + (n + 1) = 2n + 1.$$

Thus min $\deg_a F(L) = -1 - 2n$, and since $w(D) = -2n$, we have

$$\min \deg_a \Lambda(D) = -w(D) + \min \deg_a F(D) = -(-2n) - 1 - 2n = -1.$$

That $\max \deg_z[\Lambda(D)]_{a^{-1}} = 1$ can also be obtained by direct calculation.

If $\overline{k}$ is not of these special types, then $k_l \geq 3$ for some $1 \leq l \leq n$. We resolve a positive crossing in $\sigma_j^{k_l}$. We have (with $D = D_+$)

$$\Lambda(D_+) = z\Lambda(D_0) + z\Lambda(D_\infty) - \Lambda(D_-). \qquad (5.15)$$

Here D_0 has $\tilde{w}$ by one less and comes with a z-factor, so it is enough to show that $\Lambda(D_-)$ and $\Lambda(D_\infty)$ do not contribute.

If $k_l > 3$, then D_- is of the required form and has $\tilde{w}$ by two less, so by induction $\Lambda(D_-)$ has too small z-degree in $[\Lambda]_{a^{-1}}$.

Now consider $k_l = 3$. As in the proof of Lemma 6.2 of [St4] we can move by braid relations (regular isotopy) a σ_1 in $\sigma_1\sigma_2\sigma_1$ until we get a $\sigma_i\sigma_i^{-1}$ (and then cancel). Here we replaced σ_i^{-1} by δ_i, so that such σ_i right before a δ_i becomes a kink, which is negative. By repeating this transformation, we obtain a form that has no $\sigma_i\sigma_{i\pm1}\sigma_i$ (i.e., $k_i > 1$ for $1 < i < n$) and a certain non-zero number of negative kinks collected at both ends. The negative kinks shift the degree in a of Λ up, so by induction D_- has no contribution to $[\Lambda]_{a^{-1}}$.

It remains to deal with the term of D_∞ in (5.15). It is the connected sum $D_{[n_1]} \# D_{[n_2]}$ with $n_1 + n_2 = n - 1$, and with $k_l - 1 > 1$ negative kinks. So by induction $\min \deg_a \Lambda(D_\infty) = k_l - 1 + (-1) + (-1) > -1$, and $\Lambda(D_\infty)$ gives no contribution to $[\Lambda(D_+)]_{a^{-1}}$.   □

Theorem 5.5 *If a 3-braid link is positive, then it is the closure of a positive or. almost positive 3-braid. Along these links the non-fibered ones are exactly the $(1, p, q, r)$-pretzel links.*

Proof Consider the first claim. Since L is positive, it is strongly quasi-positive, and so has a positive 3-braid band representation β by Theorem 1.1. So β is of Xu's form R or $(21)^k R$ (with $k > 0$) up to extensions. If β contains $\sigma_1\sigma_2\sigma_1$ or $\sigma_2\sigma_1\sigma_2$, we can reduce it as before, until (a) it becomes positive, or (b) it still has a positive band representation, but it does not contain $\sigma_1\sigma_2\sigma_1$ or $\sigma_2\sigma_1\sigma_2$. In case (a) we are done. In case (b) we observe that β is of the form (5.14) with $n_i \geq 3$, apply Lemma 5.8, and conclude that $l \leq 1$.

It remains to argue which links are not fibered. Positive braids are always fibered, and by direct observation for an almost positive braid we have Xu's form R or $(21)^k R$ depending on whether in (5.14) (with $l = 1$) we have $n_1 = 3$ or $n_1 > 3$. We proved in Theorem 3.3 that the forms $(21)^k R$ (for $k > 0$) give fibered closure links. For $n_1 = 3$ we have as before the $(1, p, q, r)$-pretzel links.   □

Remark 5.2 Unfortunately, we cannot completely determine which of the almost positive braid representations with $l = 1$ and $n_1 > 3$ in (5.14) gives positive links. For example, the Perko knot 10_{161} has such a representation, and it is positive. The

Perko move (see [HTW]), turning the closed braid diagram of 10_{161} into a positive diagram, applies for more general examples. However, some knots, like 14_{46862}, do not seem subjectable to this or similar moves, and their positivity status remains unclear at this point.

Similarly, one would hope to prove that among these links none is a positive braid link (and this way to obtain a different, but much more insightful proof of Theorem 1.3). Using the polynomials, one can exclude certain families, for example all links of $n_1 = 5$, but a complete argument again does not seem possible.

Chapter 6
Studying Alternating Links by Braid Index

The combination of the identity (3.2) and the skein-Jones substitution (2.5) was already used in Section 4.2 to translate the determination of the 3-braid link genus from P to V. A similar line of thought will now enable us to extend the other main result in [St2], the description of alternating links of braid index 3. This result was motivated by the work of Murasugi [Mu], and Birman's problem in [Mo2] how to relate braid representations and diagrammatic properties of links. We will see how via (2.5) and the famous Kauffman–Murasugi–Thistlethwaite theorem [Ka2, Mu3, Th2] the Jones polynomial enters in a new way into the braid representation picture. The argument will lead to the braid index 3 result surprisingly easily, and then also to the classification for braid index 4 (which seems out of scope with the methods in [St2] alone). We also obtain a good description of the general (braid index) case.

Our starting point is the following general result concerning the MFW-bound (2.11). A diagram is called *special* if it has no separating Seifert circles; see [Cr]. The number of Seifert circles of D is denoted by $s(D)$.

Theorem 6.1 *Assume L is a non-trivial non-split alternating link, and $MFW(L) = k$. Then an alternating (reduced) diagram D of L has $s(D) \leq 2k - 2$ Seifert circles, and equality holds only if D is special.*

Proof As L is non-trivial and non-split, we have $1 - \chi(L) > 0$. It is also well-known, that $\max \deg_z P(L) = 1 - \chi(L)$ (see [Cr]). Now $\text{span}_v P(L) \leq 2k - 2$ by assumption. Under the substitution in (2.11) this translates to

$$\text{span} V(L) \leq 1 - \chi(L) + 2k - 2. \tag{6.1}$$

On the other hand, by [Ka2, Mu3, Th2], $\text{span} V(L) = c(D)$, and also $1 - \chi(L) = c(D) - s(D) + 1$. So

$$c(D) \leq c(D) - s(D) + 1 + 2k - 2, \tag{6.2}$$

© The Author(s) 2017

A. Stoimenow, *Properties of Closed 3-Braids and Braid Representations of Links*,
SpringerBriefs in Mathematics, https://doi.org/10.1007/978-3-319-68149-8_6

and then $s(D) \leq 2k - 1$. Now if this is an equality, then so is (6.1). Then one easily sees that $P(L)$ must have non-zero coefficients in both monomials $z^{\max \deg_z P} v^{\max \deg_v P}$ and $z^{\max \deg_z P} v^{\min \deg_v P}$. Now under the substitution in (3.2), both these monomials give a non-cancelling contribution, and one of them is not in $(v-)$degree 0, so the identity (3.2) cannot hold. Moreover, if (6.1) fails just by one, then still one of the two coefficients must be non-zero. In order its (non-cancelling) contribution on the left of (3.2) to be in degree 0, we see that either $\max \deg_z P = \min \deg_v P$, or $\max \deg_z P = -\max \deg_v P$. Using [Cr], one concludes then that these are precisely the cases of a (positively or negatively) special alternating link.
$\qquad\square$

Using [Y] and [Vo] we have some simple estimate on the unsharpness of MFW for alternating links.

Corollary 6.1 *For an alternating non-trivial non-split link L, we have* $b(L) \leq 2MFW(L) - 2$. $\qquad\square$

Of course, for many (in particular alternating) links $b(L) = MFW(L)$ (that is, MFW is exact), or at least $b - MFW$ is small. So the above estimate should be considered as a worst-case-analysis. Even if not strikingly sharp, it is still far from trivial, in view of what we already know can occur for non-alternating knots. Namely, using the construction in [K2] and the work in [BM3] (see Remark 6.1 below), one can find sequences of knots (K_i) for which MFW (in fact the full P polynomial) is constant, but $b(K_i) \to \infty$.

Another observation is that one can now extend the outcome of the work in [SV] by replacing crossing number of an alternating knot by its braid index.

Corollary 6.2 *Let $n_{g,b}$ denote the number of alternating knots of braid index b and genus g. Then $n_{g,b}$ is finite. Moreover, for g fixed, we have that $\lim_{b \to \infty} n_{g,b}/b^{6g-4} = C_g$ is a constant.*

Proof The finiteness of $n_{g,b}$ follows by (6.2). When $\chi(L)$ is fixed, and $c(L) \to \infty$, one easily sees from (6.2) that $MFW(L)$ behaves asymptotically (up to an $O(1)$, i.e. bounded, term) at least like $c(D)/2$. Now Ohyama's result [Oh] implies that $b(L)$ (and $MFW(L)$ as well) behave asymptotically exactly as $c(L)/2$. So from [SV] we have the result. $\qquad\square$

Remark 6.1 Of course we could also gain, as in [SV], an estimate on the C_g and its asymptotics for $g \to \infty$, it would just multiply by 2^{6g-4}. One should also note that the finiteness of $n_{g,b}$, which one sees from (6.2), is not necessarily clear a priori. In fact, however, Birman-Menasco proved [BM3] that $n_{g,b}$ is a finite number even for *general* (i.e., without restriction to alternating) knots. Their methods seem, though, quite unhelpful to estimate these numbers properly.

Theorem 6.1 immediately leads to the first slight sharpening of the description of alternating links of braid index 3 in [St2]. (The case of $MFW = 2$ is even more obvious, and omitted.)

Corollary 6.3 *An alternating link has $MFW = 3$ if and only if it has braid index 3.*

Proof If $MFW(L) \leq 3$, then the alternating diagram has at most 4 Seifert circles, and exactly 4 only if it is special. Apart from connected sums (which are easily handled), we obtain the diagrams of closed alternating 3-braids and the (p, q, r, s)-pretzel diagrams. By direct calculation of P we saw in [St2] that if $\min(p, q, r, s) \geq 2$, then $MFW \geq 4$, and that otherwise the pretzel link has braid index 3.   □

The case of 4-braids is now not too much more difficult.

Theorem 6.2 *Let L be a prime non-split alternating link. The following 3 conditions are equivalent:*

1. *$MFW(L) = 4$*
2. *$b(L) = 4$*
3. *L is one of the links, whose reduced alternating diagrams are described (up to mirror images) as follows:*

   a. *The Murasugi (or connected) sum of three $(2, n_i)$-torus links (with $|n_i| > 1$),*
   b. *The Murasugi (or connected) sum of a $(2, n)$-torus link with a (p, q, r, s)-pretzel link, with one of p, q, r, s equal to 1, or its mirror image, or*
   c. *a special diagram whose Seifert graph (see Section 2.1) is as shown in Figure 6.1.*

Proof $2 \Longrightarrow 1$. This follows from Corollary 6.3.

$3 \Longrightarrow 2$. That the links in condition 3 have braid index at least 4 follows from the description of the links with $b(L) \leq 3$ in [St2] (which also comes out of the proof of corollary 6.3). It is also not too hard to check that for all these links $b = 4$ by exhibiting a diagram D with $s(D) = 4$. The most systematic way seems to apply the graph index inequality of Murasugi–Przytycki [MP2] (see also [Oh]).

$1 \Longrightarrow 3$. By applying Theorem 6.1, we need to deal with non-special diagrams of at most 5 and special diagrams of at most 6 Seifert circles.

First consider the non-special diagrams.

For the fibered links (the reduced Seifert graph is a tree), Murasugi's result [Mu] leads directly to case 3a. For non-fibered links, the Seifert graph must have a cycle, which must be of length at least 4 (the Seifert graph is bipartite). Then the only option that remains is case 3b. We must still argue why one of p, q, r, s must be ± 1. This can be done using Murasugi–Przytycki's work, but one easily sees it also by a direct skein theoretic argument, which we explain.

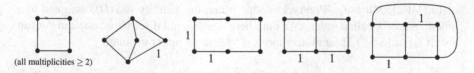

Fig. 6.1 The reduced Seifert graphs of the alternating diagrams of special alternating links of braid index 4. Simple edges have their multiplicity (1) attached, and the other edges (of multiplicity one or more) are unlabelled. In the first graph, exceptionally, all edges have multiplicity at least 2

Let $L(p, q, r, s)$ be the corresponding link and $P(p, q, r, s)$ its skein polynomial. Look first at $p = 0$, $q, r, s \geq 2$. Then $L(p, q, r, s)$ a connected sum of two $(2, n)$-torus links, and a closed alternating 3-braid. So $MFW(L(p, q, r, s)) = 5$. Now since the case $p = 1$, $q, r, s \geq 2$ has $MFW(L(1, q, r, s)) \leq b(L(1, q, r, s)) = 4$, the skein relation (2.3) easily shows that the maximal degree coefficient max cf$_v P(p, q, r, s)$ of v in P (which is a polynomial in z) for $p = 2$ is inherited from $p = 0$. Then further applications of (2.3) show that the z-degree of max cf$_v P(p, q, r, s)$ increases with p, so in particular this term never vanishes, and so $MFW(L(p, q, r, s)) = 5$.

Now consider the special diagrams. For them one considers the Seifert graph, and needs to write down all bipartite planar graphs on at most 6 vertices, which have no cut vertex. Since the diagram D is special, the placement of multiple copies of an edge gives diagrams equivalent up to flypes, so it is enough to consider simple graphs (the reduced Seifert graph), and have the multiplicities of an edge written as its label. The graphs can be easily compiled using the observation that they must contain a cycle of length 4 or 6; see Figure 6.1. By direct inspection we see that the edge multiplicities must be as specified in the figure. (In fact if an edge is multiple it turns out irrelevant what its multiplicity is, so in this case we just omit the label.) We rule out the remaining multiplicities by a skein theoretic calculation, similar to the one explained for case 3b.

Let w.l.o.g. (up to mirroring) D be positive. For each edge e in $\Gamma(D)$ of variable multiplicity $\geq i$ (where $i = 1, 2$) we calculate the skein polynomial of the diagram that corresponds to Γ for multiplicities i and $i + 1$ of e. That is, if $\Gamma(D)$ has l edges of variable multiplicity, we have 2^l polynomials to calculate. Then we check for each such set of 2^l polynomials that $Q = P_{v^{9-\chi(D)}}$ is non-zero, and max deg$_z P$ − max deg$_z Q$ as well as max cf$_z Q$ is constant within this set of 2^l polynomials. Then by (2.3) this property is inherited to diagrams D whose $\Gamma(D)$ have edges of higher multiplicity, and in particular $MFW \geq 5$.

There is one more graph,

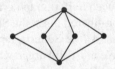

not included in Figure 6.1. In that case (by the method we just explained we verify that) $MFW \geq 5$ for all non-zero edge multiplicities.   □

In [MP2], Murasugi–Przytycki define a certain quantity ind(D), assigned to a link diagram D, called *index*. (We omit here the detailed discussion; one can consult also [Oh] or [St5].) Their motivation was to give an upper estimate

$$b(L) \leq s(D) - \text{ind}(D) \tag{6.3}$$

for the braid index of the underlying link L. Their origin of (6.3) consists in an appropriate move (see Figure 8.2 in [MP]) which reduces the number of Seifert circles of the diagram. Murasugi–Przytycki conjectured that for an alternating diagram D, the inequality (6.3) is exact. This conjecture is also confirmed for alternating links up to braid index 4.

Based on the Murasugi–Przytycki procedure, we can re-enter the Bennequin surface topic.

Corollary 6.4 *Alternating links of braid index at most 4 carry a Bennequin surface on a minimal string braid.*

Proof In [St5] it was explained how to apply restrictedly the Seifert circle reduction move of Murasugi–Przytycki (and then those of Yamada [Y]) so as to obtain a braided surface. This modified reduction is easily checked to lead to the minimal number of strings for the links in question. □

Corollary 6.5 *Let L be an alternating link of braid index 4. Then $|\max cf \Delta(L)| \leq 5$, and if $|\max cf \Delta(L)| > 2$, then L is special alternating.*

Proof It is well-known that $|\max cf \Delta(L)|$ is multiplicative under Murasugi sum and for a special diagram depends only on the reduced Seifert graph. The result follows by calculation for the specific types. □

The proof of this corollary, and the extension of the multiplicativity of $\max cf \Delta$ to $\max cf_z P$ for diagrammatic Murasugi sum [MP] demonstrates also the following more general principle:

Corollary 6.6 *For any given braid index, there are only finitely many values of $\max cf \Delta$ and $\max cf_z P$ among Alexander and skein polynomials of alternating links of that braid index.* □

We saw that for Δ this statement is wrong for non-alternating links even among 3-braids. On the other hand, by [St2], it is true for P, and we do not know if it remains true for (closed) braids on more strings. (One could also ask if infinitely many leading coefficients of Δ occur if $\max \deg \Delta = 1 - \chi$, but we see no deeper meaning in this question, so will not dwell further upon it here.)

The following is also worth observing. Call a subclass $\mathcal{C}'$ of a class $\mathcal{C}$ of links *generic* in $\mathcal{C}$ if

$$\lim_{n \to \infty} \frac{\#\{L \in \mathcal{C}' : c(L) = n\}}{\#\{L \in \mathcal{C} : c(L) = n\}} = 1.$$

Corollary 6.7 *The number of special alternating links of given braid index grows polynomially in the crossing number. In particular, a generic alternating link of given braid index is not special alternating.*

Proof A special alternating link is determined by the Seifert (= checkerboard) graph of its alternating diagram. The number of such graphs with a fixed number of vertices grows polynomially in the number of edges. The second claim in the corollary follows because it is easy to see that in contrast the number of non-special alternating links grows exponentially (due to exponentially many, in the crossing number, ways to perform the Murasugi sum at a separating Seifert circle of the alternating diagram). □

Remark 6.2 Note that in contrast we showed in [SV] that a generic alternating knot (and the case of links is analogous) of given *genus is* special alternating. This shows from yet another point of view the opposition between genus and braid index.

Another immediate and useful consequence of Theorem 6.2 and the preceding remarks is

Corollary 6.8 *If an alternating link L has* $MFW(L) \leq 4$ *(in particular if* $b(L) \leq 4$*), then the MFW inequality is exact (i.e., an equality) for L.*   □

This gives a nice complement to the MFW exactness results in [Mu, MP2]. (As another such amplification, we proved the case of *knots* and genus ≤ 4 in [St5].)

Note that at MFW bound 5 we hit already at the Murasugi–Przytycki examples [MP2] of non-exact MFW (with $b = 6$). So Corollary 6.8 is not true for $MFW \geq$ 5 or $b \geq 6$. We do not know about the case $b = 5$. However, ruling out braid index 5 for the Murasugi–Przytycki family is a serious computational problem (only two specific members were dealt with, a 4-component 15 crossing link and an 18 crossing knot; see [MP2, §19]). Already with this circumstance in mind, one cannot expect to easily extend the corollary (or Theorem 6.2) for $b = 5$ either, even if it may be true.

On the other hand, leaving these troublesome exceptions aside, the above discussion should fairly clearly explain how the general picture continues for alternating links with MFW bound 5 and more.

Chapter 7
Applications of the Representation Theory

So far the representation theory behind the skein, Jones, and Alexander polynomial was not used. We will give some applications of it now. The theory is well explained in [J]. We will use Jones' conventions, unless otherwise specified.

7.1 The Jones Conjecture

There is a conjecture stating that a minimal string braid representation has a unique exponent sum. The conjecture is often attributed to Jones, who speculated on it at least for knots and in its weaker form (as given below; see 357 1.-6 of his paper). There is also an equally popular stronger version of the conjecture, which asserts the range of exponent sum for arbitrary braid representations. (Both versions are recorded in part (a) and (b) resp. of Conjecture 1.1 in [MTr].)

Conjecture 7.1 (Jones' Conjecture)

1) (weaker version) If $\beta, \beta' \in B_n$ satisfy $\hat{\beta} = \hat{\beta}' = L$ and $n = b(L)$, then $[\beta] = [\beta'] =: w_{\min}(L)$.
2) (stronger version) Part 7.1 holds, and if $\beta'' \in B_{n'}$ for $n' > n$ has $\hat{\beta}'' = L$, then

$$\left| [\beta''] - w_{\min}(L) \right| \leq n' - n. \tag{7.1}$$

It was observed in [St3] that counterexamples to the conjecture would make MFW and *all* of its cabled versions unsharp. (In [St3] the weaker version was focussed on, but the same arguments address the stronger version too.) Thus, for example, Corollary 6.8 can also be regarded as a partial solution to Jones' conjecture. Similarly, the work in Section 5.1 shows:

Corollary 7.1 *If a link L is the closure of a positive braid, and $b(L) \leq 4$, then the strong Jones conjecture is true for L.*

© The Author(s) 2017
A. Stoimenow, *Properties of Closed 3-Braids and Braid Representations of Links*,
SpringerBriefs in Mathematics, https://doi.org/10.1007/978-3-319-68149-8_7

Proof We use Theorem 5.1, with the remark that the exceptional braid words therein were shown to have braid index 4 using the 2-cabled MFW.   □

For 3-braids the weaker version was known to be true again from Birman-Menasco's classification result. We will show now the stronger version, as a consequence of the description in [St2] of 3-braid links with unsharp MFW inequality (and thus with a much simpler proof than appealing to Birman-Menasco). See also Remark 7.6.

Theorem 7.1 *The stronger version of Jones' conjecture holds for 3-braid links.*

Convention The letter Δ, with an integer subscript n, is used in Section 7.1, in deviation from other chapters, exclusively for the half-twist braid on n strings, and *not* for the Alexander polynomial. (The Alexander polynomial will not appear in Section 7.1.)

Proof of Theorem 7.1 As explained, it suffices to deal only with the 3-braid links of unsharp MFW inequality. In [St2] these links were described fully. If $\mu = \sigma_1^{6a\pm1}\sigma_2^{\mp1}$, then for the Birman dual $\beta = \mu^*$ of μ (see Definition 4.1) we have $\hat{\mu}^* \neq \hat{\mu}$ when $6a \pm 1 > 6$, but $K = \hat{\beta}$ has the skein polynomial of the $(2, 6a \pm 1)$-torus knot $\hat{\mu}$. In [St2], is was proved that these knots K form the full list of 3-braid links with unsharp MFW inequality.

Consider first the knots $K = \hat{\beta}$ with $\beta = \mu^*$ for $\mu = \sigma_1^{6a+1}\sigma_2^{-1}$ and $a > 0$. It is easy to bring β into Xu's normal form, and to observe that it has exactly one negative band. This exhibits a Seifert ribbon (see [Ru]) of smaller genus than $g(K)$, i.e. $g_s(K) < g(K)$ (see also the proof of Theorem 1.1). Now from Xu's form one sees $w_{\min}(K) = [\beta] = 2g(K)$. Take this for a moment as a definition of $w_{\min}(K)$; this writing will be justified when we show that the writhe is unique. (One could also quote Birman–Menasco here, but the complexity of their argument is unnecessary.) So

$$1 - \chi_s(L) < w_{\min}(L). \tag{7.2}$$

Using (2.13), we see that if (7.1) is violated for some $\beta'' \in B_{n'}$, then $[\beta''] = w_{\min}(L) + 2 + (n' - 3)$. However, by the Rudolph-Bennequin inequality [Ru2] (see also [St7]), we would have then

$$1 - \chi_s(L) \geq [\beta''] - n' + 1 = w_{\min}(L),$$

in contradiction to (7.2).

The effort focusses on the other braids $\beta = \mu^*$ for $\mu = \sigma_1^{6a-1}\sigma_2$. In this case we consider $!K$, whose Xu form has only negative bands. The inequalities (2.13) show that we must rule out a braid representation $\beta'' \in B_{n'}$ of $!K$ with $[\beta''] = w_{\min}(!K) + 2 + (n' - 3)$, where (we set in the same a posteriori to be justified manner) $w_{\min}(!K) = -2g(K) - 2$.

We prohibit β'' by evaluating the 2-cable skein polynomial $P(\hat{\gamma}_a)$ of $!K$ for

$$\gamma_a = [4354] \cdot [2132]^{6a-1} \, (\Delta_6^2)^{-2a} \in B_6, \qquad (7.3)$$

with $a > 1$ and $\Delta_6^2 = [12345]^6$ generating the center of B_6, and showing that the polynomial has non-zero terms in v-degree $[\gamma_a] - 5$ or $[\gamma_a] - 3$. This may appear a banality, but in fact requires a substantial use of Jones' work in [J].

Consider the more general 6-braids

$$\beta_{k,l} := [4354] \cdot [2132]^k \, \Delta_6^{2l} \, .$$

(We are of course only interested in the special case $k = 6a - 1, l = -2a$ and $a > 1$, but it is useful to treat the 2-parameter family first.)

The skein polynomial of $\hat{\beta}_{k,l}$ can be evaluated using the representation theory in [J]. We adopt the convention that *all references* to lemmas, page and paragraph numbers, equations of the form $(x.y)$, etc. for the rest of this proof are understood to be *to Jones' paper*, unless noted otherwise.

Jones' version of the skein polynomial,

$$X(q, \lambda) = P(v, z) \quad \text{with } v = \sqrt{\lambda q} \text{ and } z = \sqrt{q} - 1/\sqrt{q},$$

can be evaluated on a closed n-braid β by a weighted sum, with weights W_Y in λ and q, of traces (in q) of irreducible representations (irreps) π_Y of B_n, indexed by Young diagrams (or tableaux[1]) Y, or equivalently by partitions of n. (A partition of n is a tuple $(n_1, \ldots, n_k)$ with $n_k > 0, n_i \geq n_{i+1}$ and $\sum n_i = n$.) We may identify the Young diagram with its partition, counting partitions in horizontal rows. For example, the partition $(4) = \boxed{}$ means one row (trivial representation), while $(1111) = \boxed{}$ means one column (parity representation); thus, $\pi_{3,1} = \pi_{\boxed{}}$. Since the calculation of W_Y was given in [J], we will not repeat it in detail. We will also deal (mostly) with Y where the calculation of π_Y is explained in [J].

Using Definition 6.1 (of [J]), (5.5) and the formula for $W_Y(q, \lambda)$ from p. 347 top, one has for a 6-braid β of exponent sum $e = [\beta]$,

$$X_{\hat{\beta}}(q, \lambda) = -\sqrt{\lambda}^{e-5} \sum_{Y \vdash 6} \widetilde{W}_Y(q, \lambda) \operatorname{tr} \pi_Y(\beta) \, . \qquad (7.4)$$

Here $\widetilde{W}_Y$ denote the slightly rescaled weights

$$\widetilde{W}_Y(q, \lambda) = \frac{R_Y(q, \lambda)}{Q_Y(q)} \cdot \frac{1 - q}{1 - \lambda q},$$

---

[1] Here consistently "tableau" is used as a synonym for (Young) diagram, i.e. with no additional information attached to it.

with R_Y being specified on p. 347 after Figure 5.6, and $Q_Y(q)$ being the hook length product term from p. 346 middle. The symbol "⊢" is taken from partition theorists and is used to mean here that Y is a Young tableau of 6 boxes (or equivalently a partition of 6). In order to avoid denominators it is useful to multiply (7.4) by $(1 - q^2) \cdot \ldots \cdot (1 - q^6)$, so

$$- (1 - q^2) \cdot \ldots \cdot (1 - q^6) X_{\hat{\beta}} = \sqrt{\lambda}^{e-5} \sum_{Y \vdash 6} \widehat{W}_Y(q, \lambda) \operatorname{tr} \pi_Y(\beta), \qquad (7.5)$$

with

$$\widehat{W}_Y(q, \lambda) = \frac{R_Y(q, \lambda)}{1 - \lambda q} \cdot \frac{(1 - q)(1 - q^2) \cdot \ldots \cdot (1 - q^6)}{Q_Y(q)}$$

becoming a Laurent polynomial in λ and q.

Note that π_Y are representations that involve only the variable q, not λ. The obvious question is how to evaluate their traces.

Consider first

$$\beta_k = \beta_{k,0} = [4354] [2312]^k.$$

There are 11 Young diagrams Y of six boxes. Now $\beta_k \in B_4 \subset B_6$, and according to p. 340 top, the restriction of π_Y to B_4 splits into a direct sum of representations $\pi_{Y'}$ indexed by 4-box Young diagrams Y', which we name for shorthand

$$A = \quad B = \quad C = \quad D = \quad E = \quad . \qquad (7.6)$$

The representations $\pi_{Y'}$ are clarified completely in §8 (see also Section 7.2 in *this* monograph below). Denote by ψ_{n-1} the (reduced) $(n - 1)$-dimensional representation of B_n; see §2. Let us also write -1 for the parity representation $(-1)^{[\cdot]}$, and $-\rho = -1 \otimes \rho$ for the direct (tensor, or Kronecker) product of a representation ρ with the parity. Then $\pi_A = -1$, π_E is given by $q^{[\cdot]}$, $\pi_B = -\psi_3$ (Note 5.7; the sign disappears here, though, because for us always $e = [\beta]$ is even), $\pi_D = - \bigwedge^2 \psi_3$ and $\pi_E = \pi_{\square} \circ \bar{}$, with $\bar{}$ being the homomorphism $B_4 \to B_3$ given by $\overline{\sigma_{1,2,3}} = \sigma_{1,2,1}$ (p. 355). Also $\pi_{\square} = -\psi_2$.

Let us write below $\overline{Y}$ for the *transposed* (or *dual*) Young diagram to Y, given by exchanging rows and columns. For example, for $Y = \square$, we have $\overline{Y} = \square$. The relation between π_Y and $\pi_{\overline{Y}}$ is given in Note 4.6.

Now $\psi_3(\alpha)$ and $\psi_2(\bar{\alpha})$ for

$$\alpha = [2132]$$

are easy to calculate (see §2). We find former's eigenvalues to be $\pm t$ and $-t^2$, and latter's t and t^3. In particular, both matrices are diagonalizable (because the eigenvalues are distinct for generic t). Setting $t = q$ as in Note 5.7, we have the following table of eigenvalues of the $\pi_{Y'}(\alpha)$

A	B	C	D	E
1	$-q$	q^3	$-q^3$	q^4
	q	q	q^3	
	$-q^2$		$-q^2$	

i	1	2	3	4	5	6	7
δ_i	1	q^4	q	$-q$	$-q^2$	q^3	$-q^3$

Let us number the seven possible eigenvalues of $\pi_Y'(\alpha)$ by δ_i as shown on the right.

Thus $\pi_Y(\alpha)$ are all diagonalizable, with eigenvalues δ_i. The multiplicities of $\pi_{Y'}$ in π_Y are also easy to calculate. They are the number of descending paths from Y' to Y in Figure 3.3 (continued one more row to the bottom). Table 7.1 shows the multiplicities of $\pi_{Y'}$ (in the shorthand of (7.6)) in π_Y and the resulting ones, which we write $M(i, Y)$, of the eigenvalues δ_i in $\pi_Y(\alpha)$.

Since with $\pi_{Y'}(\alpha)$ also all $\pi_Y(\alpha)$ are diagonalizable, we have

$$\operatorname{tr} \pi_Y(\beta_k) = \operatorname{tr} \pi_Y(\alpha^k \cdot [4354]) = \sum_{i=1}^{7} c(i, Y)\,\delta_i^k, \tag{7.7}$$

with $c(i, Y)$ given as follows. Consider $\pi_Y([4354])$ in the basis of π_Y that diagonalizes $\pi_Y(\alpha)$. Then to obtain $c(i, Y)$, sum the (j, j)-entries of the matrix of $\pi_Y([4354])$ over rows/columns j, for which the (j, j)-entry of $\pi_Y(\alpha)$ is δ_i.

So the problem to evaluate $\operatorname{tr} \pi_Y(\beta_k)$ transforms into the one to determine $c(i, Y)$. There are a priori 77 of those, given by combining 7 eigenvalues δ_i with 11 Young diagrams Y. However, one immediately notes that clearly $c(i, Y) = 0$ when $M(i, Y) = 0$. This leaves 45 of the 77 values.

If we can calculate $\operatorname{tr} \pi_Y$ for general 6-braids, then we can use (7.7) as a linear equation for $c(i, Y)$. If $\pi_Y(\alpha)$ has l_Y different eigenvalues, we can determine $c(i, Y)$ for that Y by calculating the l.h.s. of (7.7) for l_Y different values of k and solving for $c(i, Y)$. (Always $l_Y \le 6$, as evident from Table 7.1.)

In case Y is one of the one-hook diagrams like ⊞, then π_Y is by p. 354 bottom (a tensor product of a parity, which disappears at even exponent sum, with) an exterior power of ψ_5. Thus, as in §7 (of Jones' paper), one can evaluate $\operatorname{tr} \pi_Y(\beta_k)$ from the characteristic polynomial of the Burau matrix $\psi_5(\beta_k)$. We determined this way the corresponding $c(i, Y)$, and verified them with a few extra values of k in (7.7). While it is clear that all $c(i, Y)$ should be rational expressions in q, I expected them to be in fact Laurent polynomials. So I was a bit startled by the denominators $1 + q^2$. However, according to p. 343 top, the Hecke algebra may degenerate at roots of unity q, which justifies at least cyclotomic polynomials as denominators.

Table 7.1 This table displays the Young tableaux Y of six boxes, the decomposition into $\pi_{YY'}$ for 4-box Young tableaux Y' of the sub-representation of B_4 in π_Y (writing Y' for $\pi_{Y'}$), the multiplicities $M(i, Y)$, the quantities d ($= \dim \pi_Y$) and r (rank of the idempotent e_1) occurring in Jones' lemma 9.3

Y / Y'	A	$2A+B$	$A+2B+C$	$B+C$	$A+2B+D$	$2B+2C+2D$	$C+D$	$B+2D+E$	$C+2D+E$	$D+2E$	E
1	1	2			1						
q^4		1	1								
q		1	3	2	2	4	1	1	1	2	1
$-q$		1	2	1	2	2		1	1		
$-q^2$		1	2	1	3	4	1	3	2	1	
q^3			1	1	1	4	2	2	3	1	1
$-q^3$					1	2	1	2	2	1	1
d	1	5	9	5	10	16	5	10	9	5	1
r	0	1	3	2	4	8	3	6	6	4	1
$30r/d$	0	6	10	12	12	15	18	18	20	24	30

For $Y = $ ⊞, π_Y was written down directly on p. 362, and for its transposed (dual) Young diagram ⊟, one uses Note 4.6.

There remain three representations, for $Y = $ ⊟, ⊞⊟, and its dual ⊟, with a total of 15 unknown $c(i, Y)$. These were more complicated to find, since we knew of no way to evaluate $\operatorname{tr} \pi_Y$ directly. To help ourselves, first observe that we have, for fixed Y, the trace identities

$$\sum_{i=1}^{7} c(i, Y) = \operatorname{tr} \pi_Y([4354]) = \operatorname{tr} \pi_Y([2132]) = \sum_{i=1}^{7} M(i, Y)\delta_i , \qquad (7.8)$$

which again give a linear condition on the $c(i, Y)$. To find further identities, we used (7.5) in a "backward" manner. We calculated for small[2] k, l (k odd, $|k| \leq$ 5, $|l| \leq 1$) the polynomial on the left using Morton–Short's program [MS2]. We substituted the known $c(i, Y)$ on the right of (7.5), obtaining thus linear conditions for the yet unknown $c(i, Y)$. To determine the coefficients, it remains to understand the effect on $\operatorname{tr} \pi_Y$ of multiplying with the full twist Δ_6^2. This was, however, done also by Jones in §9, Lemma 9.3:

$$\pi_Y(\Delta_n^2) = q^{rn(n-1)/d} Id_{\pi_Y} . \qquad (7.9)$$

For given Y, the number $r = r_Y$ is calculated as in Lemma 9.1 from Figure 3.3, and $d = d_Y = \dim \pi_Y$ more easily by the hook length formula on p. 341. Call $e_Y = n(n-1)r/d$ (with $n = 6$) the exponent[3] of q on the right hand-side of (7.9). These values are given in Table 7.1. Two simple checks are $\sum_{Y \vdash n} d_Y^2 = n!$ (because the multiplicity of each irrep in the Hecke algebra equals its dimension), and that e_Y are integers (Remark on p. 358 bottom) and satisfy $e_Y + e_{\overline{Y}} = n(n-1)$ for all Y (because of Note 4.6). So (7.5) gets

$$-(1-q)\cdot \ldots \cdot(1-q^6) X(\hat{\beta}_{k,l})(q, \lambda) = \sqrt{\lambda}^{e-5} \sum_{Y \vdash 6} \widehat{W}_Y(q, \lambda) q^{l \cdot e_Y} \sum_{i=1}^{7} c(i, Y) \delta_i^k , \qquad (7.10)$$

with $e = [\beta_{k,l}] = 4 + 4k + 30l$.

Actually, each polynomial $X(\hat{\beta}_{k,l})$ gives six equations for $c(i, Y)$, because there are six relevant λ-coefficients on both hand-sides of (7.5) (in degrees $\frac{e-5}{2}, \frac{e-3}{2},$

---

[2]In the parametrization $P(v, z)$, used by Morton–Short, unlike for X, the coefficients of $P(\hat{\beta}_{k,l})$ become quickly large and produce machine size integer overflows. In particular, we could not calculate correctly polynomials for $|l| > 2$.

[3]This is not to be confused with the variable e, which we use for exponent sum of a braid, or with the idempotent e_i from Jones' lemma 9.1.

..., $\dfrac{e+5}{2}$; it is helpful to multiply again by $(1+q^2)$ to get disposed of the denominators of the known $c(i,Y)$). We have with 18 polynomials 111 equations (6 equations per polynomial plus the three relevant trace equations (7.8)). Still the resulting system was too hard to solve by computer, using MATHEMATICA™ [Wo], since its coefficients are (Laurent) polynomials in q, with dozens of terms each.

However, substituting some (rational) values of q, the system can be solved immediately. We used this to check first the rank of the matrix (i.e., which equations are linearly redundant). Again we were surprised that for 15 variables $c(i,Y)$ the rank was only 14. This, however, can be explained from our restraint to odd k (which we chose for some, purely technical, component number concerns in the calculation with Morton-Short's program). Whenever two opposite eigenvalues δ_i and $\delta_{i'} = -\delta_i$ occur in $\pi_Y(\alpha)$, the equations (7.5) for odd k can detect only $c(i,Y) - c(i',Y)$. The trace equations (7.8), which involve $c(i,Y)$ and $c(i',Y)$ with the same sign, remedy the shortcoming for ⊞⊟ and its transposed diagram, but for ⊟⊞ we have two pairs (i,i') of opposite eigenvalues, so we still lose one dimension. This is, however, not really a problem, because in the braids $\gamma_a = \beta_{6a-1,-2a}$ of (7.3) we need for our proof, $k = 6a - 1$ is always odd, so we need only $c(i,Y) - c(i',Y)$ to evaluate the polynomial in (7.5).

We used the special evaluations to select equations that give the full matrix rank, and to guess the formula for general q for some of the $c(i,Y)$. (The ones we already found suggest that these formulas should not be so complicated.) Substituting these (yet potential) solutions, too, gives an even simpler linear system for the still unknown $c(i,Y)$, which then could be solved in q.

Since we know that our matrix has rank 14, it is enough to check our solution (up to the one-dimensional ambiguity, which will disappear in (7.5)) with the 111 equations we have. The result was confirmed, and is shown in Table 7.2.

With all $c(i,Y)$ determined, the main work is done. So far we can evaluate $X(\hat{\beta}_{k,l})$ for odd k. Multiplying (7.10) by $Z = (1+q)(1+q^2)(1+q+q^2)$ to clear all denominators in the $c(i,Y)$, normalizing and taking coefficients in λ^m for $m = 0, \ldots, 5$, we have (with e_Y being the exponents of q in (7.9) and $e = [\beta_{k,l}]$)

$$\left[-\frac{1}{\sqrt{\lambda}^{e-5}} \cdot Z \cdot (1-q) \cdot \ldots \cdot (1-q^6) X_{\hat{\beta}_{k,l}}(q,\lambda) \right]_{\lambda^m} = \qquad (7.11)$$

$$= \sum_{Y \vdash 6} \sum_{i=1}^{7} (Z \cdot c(i,Y)) \, \delta_i^k \, q^{l \cdot e_Y} \left[\widehat{W}_Y(q,\lambda) \right]_{\lambda^m} .$$

Recall that for our proof it is enough to show that this term becomes non-zero for $m = 0$ or $m = 1$, when $k = 6a - 1$ and $l = -2a$ for an integer $a > 1$ (and $\beta_{k,l} = \gamma_a$ in (7.3)). Now, for odd k (and fixed Y), we can group the sum over 7 terms $Z \cdot c(i,Y)$ into five terms $\tilde{c}(i,Y)$ accounting for $\delta_3 = -\delta_4$ and $\delta_6 = -\delta_7$, and thus excluding $i = 4, 7$. So the above sum in (7.11) becomes

Table 7.2 This table shows the values $c(i, Y)$ of (7.7) for Young tableaux Y of six boxes and the seven possible eigenvalues δ_i of $\pi_Y(\alpha)$

Y (δ_i)	[Y₁]	[Y₂]	[Y₃]	[Y₄]	[Y₅]	[Y₆]	[Y₇]	[Y₈]
1	1	$-\left(\dfrac{q^4}{1+q^2}\right)$	$\dfrac{q^7}{(1+q^2)(1+q+q^2)}$	0	$q+\dfrac{q^2}{2}$	0	0	0
q^4	0	$-\left(\dfrac{q^2}{1+q^2}\right)$	$\dfrac{q}{(1+q^2)(1+q+q^2)}$	0	$\dfrac{q^2}{2}$	0	0	q^4
q	0	0	$\dfrac{q^6}{1+q+q^2}$	$\dfrac{q^6}{2+2q^2}$	$-q^2$	$\dfrac{q(1-4q+q^2)}{2} - c\!\left(6,\blacksquare\right)$	$\dfrac{q^2+2q^4}{-2-2q^2}$	0
$-q$	0	0	0	$\dfrac{q^6}{2+2q^2}$	$-q^2$	$\dfrac{(1-q)^2 q}{2} - \dfrac{q^2}{1+q} - c\!\left(6,\blacksquare\right)$	$\dfrac{q^2+2q^4}{-2-2q^2}$	0
$-q^2$	0	$\dfrac{q^6}{1+q^2}$	$\dfrac{q^3(1-q+q^2)}{1+q^2}$	$-q^2$	$-q^2$	$q\,(1+(-1+q)\,q)$	$-q^2$	0
q^3	0	$\dfrac{q^4}{2}$	$\dfrac{(-1+q)q^2}{2}$	$-\dfrac{q^2(2+q^2)}{2(1+q^2)}$	$\dfrac{q}{2}+q^3$	$c\!\left(6,\blacksquare\right)$	$\dfrac{1}{2+2q^2}$	0
$-q^3$	0	$\dfrac{q^4}{2}$	$\dfrac{(-1+q)q^2}{2}$	$-\dfrac{q^2(2+q^2)}{2(1+q^2)}$	$\dfrac{q}{2}$	$\dfrac{q^2}{1+q}+c\!\left(6,\blacksquare\right)$	$\dfrac{1}{2+2q^2}$	0

$$\sum_{Y \vdash 6} \sum_{\substack{i=1 \\ i \neq 4}}^{6} \tilde{c}(i, Y) \delta_i^k q^{l \cdot ey} \left[\widehat{W}_Y(q, \lambda) \right]_{\lambda^m}. \tag{7.12}$$

(Now in the $\tilde{c}(i, Y)$, the 1-degree ambiguity of $c(i, Y)$ for $Y = \boxplus$, as explained above, cancels out.)

Among the 55 possible (i, Y) (with $i \neq 4, 7$), only 21 of the $\tilde{c}(i, Y)$ are non-zero. It turns out that for $m = 0$, when $k = 6a - 1$, $l = -2a$ and $a > 1$, there is a unique term among the 21 summands in (7.12) whose minimal degree in q is the smallest (it is $14 - 36a$). Thus $[P(\hat{\beta}_{k,l})]_{v^{e-5}} \neq 0$. We calculated the polynomial with Morton-Short's program for $a = 2$ (where the calculation was still feasible), and it confirmed that all six v-terms appear. We also calculated that for $m = 0$ and $a = 0, 1$ there are two terms of smallest minimal degree. This better ought to be so, because in that case $\hat{\gamma}_a$ are just a 2-cable of the unknot and $!5_1$, resp., and the coefficients for $m = 0, 1$ in (7.12) must be 0, which we checked once more separately. (Also we found that for $m = 1$ there are two terms of smallest minimal degree in (7.12) for all $a \geq 0$.) With this the proof of Theorem 7.1 is complete.

Let us finally say that the computer part of the calculation owed a lot to the use of MATHEMATICA. While, if done properly, it could be carried out in a few minutes, it required a week of work to find the way of skillfully programming MATHEMATICA to do all the separate steps in an efficient way.   □

Remark 7.1 Note that one could handle the cases $\mu = \sigma_1^{6a+1} \sigma_2^{-1}$ from the beginning of the proof also using the representation theoretic argument, by looking at $m = 4, 5$ for $a < 0$. We waived on this investigation, though, since the proof for $a > 0$ was laborious enough.

7.2   Unitarity of the Burau Representation

In the following q and t are unit norm complex numbers. We define $\arg(e^{is}) := s \mod 2\pi$ for $s \in \mathbb{R}$. We continue using the formalism of Young tableaux, the representations π_Y and the notations $X(q, \lambda)$ and W_Y of the proof of Theorem 7.1.

For the Alexander polynomial Δ (we resume the notational convention from before Section 7.1), as well as for 3- and 4-braids, the representations π_Y are given by Burau representations. We note (again, and more explicitly) the following descriptions of π_Y in terms of the Burau representation ψ_n given in [J]. Again the indexing is chosen so that ψ_{n-1} is the reduced $(n-1)$-dimensional representation of B_n, and by $-\rho$ we denote the direct product of ρ with the parity representation.

As before, e stands for the exponent sum of a braid β. For 3-braids we have the following properties (with reference to the explanation in [J]):

1. $\pi_{\square}(\beta)(q) = (-1)^e \psi_2(\beta)(q) = q^e \psi_2(\beta^{-1})(q)$ (because of row-column symmetry; see Note 4.6).
2. $\pi_{\square}(\beta) = (-1)^e$ and $\pi_{\square\square\square}(\beta) = q^e$ (Note 4.7)

For 4-braids we have:

1. $\pi_{\square} = -\psi_3$. So $\pi_{\square}(\beta) = (-1)^e \psi_3(\beta)$. (Note 5.7)
2. $\pi_{\square\square}(\beta)(q) = q^e \psi_3(\beta^{-1})(q)$ (because of row-column symmetry; see Note 4.6). Also

$$\pi_{\square\square\square} = -\pi_{\square} \wedge \pi_{\square} = -\psi_3 \wedge \psi_3 \, ,$$

where wedge denotes antisymmetric product (see p. 354 bottom).
3. $\pi_{\boxplus}(\beta) = (-1)^e \psi_2(\bar{\beta})$, where bar denotes the homomorphism from B_4 to B_3 given by $\bar{\sigma}_{1,2,3} = \sigma_{1,2,1}$. (p. 355)
4. $\pi_{\square}(\beta) = (-1)^e$ and $\pi_{\square\square\square\square}(\beta) = q^e$ (Note 4.7)

Now Squier observes in [Sq], that $\psi_i(\beta^{-1})(q)$ and $\psi_i(\beta)(q^{-1})$ are conjugate and so have the same trace. So by the self-symmetry of $\pi_{\square}$ we have $\operatorname{tr}\pi_{\square}(t), \operatorname{tr}\pi_{\boxplus}(t) \in (-t)^{e/2}\mathbb{R}$. Similarly $(-t)^{-e/2}\operatorname{tr}\pi_{\square}(t)$ and $(-t)^{-e/2}\operatorname{tr}\pi_{\square\square}(t)$ are conjugate complex numbers. These properties will be important below.

Remark 7.2 Squier uses a different convention for ψ_i from Jones. He transposes and changes sign in matrix entries with odd row-column sum (i.e., conjugates by $\operatorname{diag}(1, -1, 1, -1, \dots)$). This, however, does not affect our arguments.

The key point in arguments below is Squier's result. We write M^* for the conjugate transposed of a matrix M. (That is, $M^*_{i,j} = \overline{M}_{j,i}$.)

Theorem 7.2 (Squier [Sq]) *For any $n \geq 1$ there exist a Hermitian matrix $J = J^{[n+1]}$ and a regular matrix M, such that with $J_0 = M^*JM$ we have $\psi_n^* J_0 \psi_n = J_0$.*

In particular, J is degenerate or definite iff J_0 is so. Moreover,

$$J_{i,j} = \begin{cases} -1 & \text{if } |i - j| = 1 \\ \sqrt{t} + 1/\sqrt{t} & \text{if } i = j \\ 0 & \text{otherwise} \end{cases} .$$

It is easy to see that if $t = 1$, then J is positive definite. Now definiteness is an open condition, so for t close to 1, it is still valid. One can determine when J loses this property.

Proposition 7.1 *The Squier form $J^{[n]}$ on B_n degenerates exactly in the nth roots of unity. In particular, it is positive definite exactly when $|\arg t| < 2\pi/n$.*

Proof Denote by $J^{[n]}$ the form corresponding to n-braids, i.e. the one given by restricting J to the first $n-1$ rows and columns. It is not too hard to calculate the determinant of $J^{[n]}$. By development in the last row,

$$\det J^{[n]} = (\sqrt{t} + 1/\sqrt{t}) \det J^{[n-1]} - \det J^{[n-2]} ,$$

whence

$$\det J^{[n]} = \frac{t^n - 1}{\sqrt{t} - 1/\sqrt{t}} \cdot \frac{1}{(\sqrt{t})^n} .$$

Then the claim follows easily. To see definiteness use the (positivity of) the principal minor criterion. Since $e^{\pm 2\pi i/n}$ is a simple zero of $\det J^{[n]}$, the determinant must become negative for $|\arg t| \in (2\pi/n, 4\pi/n)$. Then applying this argument to all $n' < n$ shows that $J^{[n]}$ is not positive definite for $|\arg t| \geq 2\pi/n$. $\qquad \square$

So on the arcs of S^1 that connect the primitive nth root of unity to 1, we have that J is positive definite. Now if J is such, it can be written as $Q^* Q$, and then conjugating ψ_i by QM we obtain a $U(n-1)$-representation. This means in particular that all eigenvalues of ψ_i have unit norm. We will below derive implications of this circumstance for the link polynomials.

7.3 Norm Estimates

The Jones polynomial V can be specified, for our purposes, by $V(t) = X(t,t)$. In the following, which root of complex numbers is taken is irrelevant, important is though that it be kept fixed in subsequent calculations. By $\Re$ we denote the real part of a complex number.

Theorem 7.3 *If $|t| = 1$, $\Re t > 0$ and β is a 4-braid, then $\left| V_{\hat{\beta}}(t) \right| \leq (2\Re \sqrt{t})^3$. If β is a 3-braid and $\Re t > -1/2$, then $\left| V_{\hat{\beta}}(t) \right| \leq (2\Re \sqrt{t})^2$.*

Proof We have from [J] that if $\beta \in B_4$ with $[\beta] = e$, then

$$V_{\hat{\beta}}(t) = (-\sqrt{t})^{e-3} \left[\frac{t(1-t^3)}{1-t^2} \operatorname{tr} \psi_3 + \frac{t^2}{1+t} \operatorname{tr} \bar{\psi}_2 + \frac{1-t^5}{1-t^2} \right], \qquad (7.13)$$

where $\bar{\psi}_2$ is the composition of ψ_2 with $^- : B_4 \to B_3$. Taking norms and using that $\bar{\psi}_2$ and ψ_3 are unitary, we find

$$|V(t)| \leq 3 \left| \frac{1-t^3}{1-t^2} \right| + \frac{2}{|1+t|} + \left| \frac{1-t^5}{1-t^2} \right| . \qquad (7.14)$$

It is now a routine (but somewhat tedious) calculation to verify that the r.h.s. is equal to $(2\Re \sqrt{t})^3$ for $|t| = 1$, $\Re t > 0$.

For $\beta \in B_3$ we have similarly

$$V_{\hat{\beta}}(t) = \left(-\sqrt{t}\right)^{e-2}\left[t \cdot \mathrm{tr}\,\psi_2 + (1 + t^2)\right], \qquad (7.15)$$

and the result follows using $|\mathrm{tr}\,\psi_2| \le 2$. $\qquad\qquad\qquad\qquad\qquad\qquad\quad\square$

This theorem generalizes Jones' result [J, Proposition 15.3] for $n \le 4$, where he considers $t = e^{2\pi i/k}$, $k \ge 5$. In fact the comparison to (and established coincidence with) Jones' estimate led to the simplification of the r.h.s. of (7.14). In [St3] we noted that Jones' estimate can be better than MFW when $MFW = 3$, but by connected sum one can give an example for $MFW = 4$. Again it appears that for $\beta \in B_4$ and $t \ne e^{\pm\pi i/3}$ the set $\{|V_{\hat{\beta}}(t)|\}$ is dense in $[0, (2\Re\sqrt{t})^3]$, and similarly it is in $[0, (2\Re\sqrt{t})^2]$ for $\beta \in B_3$ and $t \ne e^{\pm\pi i/3}, e^{\pm\pi i/5}$. (See also the remarks at the end of §12 in [J].)

Conjecture 7.2 If $\beta \in B_n$ and $|t| = 1$ with $|\arg t| < 2\pi/n$, then $|V(t)| \le (2\Re\sqrt{t})^{n-1}$.

In the case of the Alexander polynomial $\Delta(t) = X(t, 1/t)$, we can say something on general braids.

Theorem 7.4 *For each $n \ge 2$, if $|t| = 1$ and $|\arg t| \le 2\pi/n$ and $\beta \in B_n$, then*

$$|\Delta_{\hat{\beta}}(t)| \le \frac{2^{n-1}|1 - t|}{|1 - t^n|}.$$

Proof J_0 is positive definite when $|t| = 1$ and $|\arg t| < 2\pi/n$. Then

$$|\Delta_{\hat{\beta}}(t)|\frac{|1 - t|}{|1 - t^n|} = |\det(1 - \psi_{n-1}(\beta))|,$$

and all eigenvalues of $1 - \psi_{n-1}$ have norm ≤ 2. The case $|\arg t| = 2\pi/n$ follows by continuity. $\qquad\qquad\qquad\qquad\qquad\qquad\qquad\qquad\qquad\qquad\qquad\qquad\quad\square$

Corollary 7.2 *For all n, k the set*

$$\left\{\Delta(\hat{\beta}) : \beta \in B_n,\ \deg\Delta \le k\right\}$$

is finite. That is, among closed braids of given number of strands only finitely many Alexander polynomials of given degree occur.

Proof Δ is determined by $\Delta(t_i)$ for k different t_i with $|t_i| = 1, 0 < \arg t_i < 2\pi/n$, by means of a linear transformation using the (regular Vandermonde) matrix $M = (t_i^j)_{i,j=1}^n$. So $||[\Delta(t)]_{i}|| \le ||M^{-1}|| \cdot ||\Delta(t_i)||$. $\qquad\qquad\qquad\qquad\qquad\qquad\quad\square$

This result should be put in contrast to the various constructions of knots with any given Alexander polynomial. For example, a recent construction of Nakamura [Na2] allows to realize the degree of the polynomial by the (actually braidzel) genus of the knot. (We were subsequently independently able to further specialize this result to canonical genus.) A different construction of Fujii [Fu] shows that knots with three bridges admit all Alexander polynomials. So the situation between braid and plat closures is completely different.

Compare also Birman-Menasco's result in [BM3], mentioned (for knots) in Remark 6.1, that there are only finitely many closed braids of given number of strands with given genus. Note that we do not claim that only finitely many closed braids of given number of strands with given Alexander polynomial (degree) occur. For 3-braids it is true, but from 5-braids on the non-faithfulness of the Burau representation should (in principle, modulo the evaluability of another invariant) make it possible to construct infinite families of links with the same (for example, trivial) polynomial. It makes some sense to ask about the status of 4-braids.

Question 7.1 Are there only finitely many closed 4-braids of given Alexander polynomial (degree)?

When working with Δ, for 3- and 4-braids we can be more explicit.

Corollary 7.3 *If $\beta \in B_4$, then*

$$\left| \Delta_{\hat{\beta}}(t) \right| \leq \frac{8\,|1-t|}{|1-t^4|}, \tag{7.16}$$

when $|t| = 1$ and $\Re t > 0$. If $\beta \in B_3$ and $\Re t > -1/2$, then

$$\left| \Delta_{\hat{\beta}}(t) \right| \leq \frac{4\,|1-t|}{|1-t^3|}.$$

$\square$

Putting $t = e^{2\pi i/5}$ in (7.16), we have $\left| \Delta(e^{2\pi i/5}) \right| \leq 8$. This improves the bound $10.47 \cdots$ in [St3] suggested to replace Jones' (incorrect) value 6.5 in [J2]. For 3-braid *knots* Jones gives in [J, Proposition 15.2] the better bound 3 when $t = i$, using the property $V(i) = \pm 1$.

Example 7.1 The simplest knots with $MFW = 4$ which can be excluded from being a 4-braid using Corollary 7.3 are 13_{8385} (where one can use $t = e^{2\pi i/9}$) and 14_{37492} (with $t = e^{2\pi i/8}$). For 3-braids we can deal with the known examples 9_{42} and 9_{49}. (This gives now an alternative proof that there are no 3-braid knots with such Alexander polynomial, as we explained in Example 4.2.)

If we know the exponent sum of β we can do better.

Proposition 7.2 *If $\beta \in B_4$, $e = [\beta]$ and t as before, then*

$$\left| (-\sqrt{t})^{e-3} \, \Delta_{\hat{\beta}}(t) \frac{1-t^4}{1-t} - 1 + (-t)^e \right| \leq 6.$$

If $\beta \in B_3$ and $\Re t > -1/2$, then

$$\left| (-\sqrt{t})^{e-2} \, \Delta_{\hat{\beta}}(t) \frac{1-t^3}{1-t} - 1 - (-t)^e \right| \leq 2 \,.$$

□

Example 7.2 Since one can determine the possible e via MFW from P, one can apply Proposition 7.2 for given P. In the 3-braid case we can exclude 10_{150} this way. The remaining two 10 crossing knots with unsharp MFW, 10_{132} and 10_{156}, fail – understandably, since they share the skein polynomials of (the closed 3-braids) 5_1 and 8_{16} resp. (See the table[4] in [J].) For 4-braids several new 14 crossing knots can be ruled out, for example 14_{21199}.

Remark 7.3 The use of P to restrict the possible values of e is usually most effective, but not indispensable. There are other conditions on e, originating from Bennequin's work [Be], that can be more applicable in certain cases where the calculation of P is tedious. Also, when t is a root of unity of order n, the tests depend only on e mod $2n$.

Remark 7.4 Note that the quantity χ in (4.5) is equal to $u \operatorname{tr} \psi_2$, where $u = \sqrt{-t} + \dfrac{1}{\sqrt{-t}}$ and ψ_2 is the Burau matrix of B_3 (see the explanation in [K]). Thus one can obtain similar estimates for values of Q on 3-braids.

7.4 Skein Polynomial

Now it is natural to look at the full 2-variable skein polynomial X. We have, as in (7.4), for $\beta \in B_4$ of exponent sum e,

$$X_{\hat{\beta}}(q, \lambda) = -\sqrt{\lambda}^{e-3} \sum_{Y \vdash 4} \operatorname{tr} \pi_Y(q) \, \widetilde{W}_Y(q, \lambda) \,,$$

where the weights $\widetilde{W}_Y$ are given in λ-coefficients by Table 7.3. (They are all polynomials in λ of degree 3, with coefficients being rational expressions in q.) Now, with given e and P, we have four equations (the coefficients in λ) in three unknowns (the traces of π_{211}, π_{31} and π_{22}; we use here the partition notation for the subscripts). However, the restriction of the matrix in Table 7.3 to the columns of π_{211}, π_{31} and π_{22} has rank 2. This means that two of the $X_i = [X]_{\lambda^{i+(e-3)/2}}$ for $0 \leq i \leq 3$ determine the other two. One could believe now to use this as a 4-braid test. However, these two relations result from the general substitutions $\lambda = 1$ and $\lambda = 1/q^2$ that turn X into the component parity count or 1. These substitutions kill all trace weights except of the trivial or parity representation, and for these representations the weights become also independent on the braid group. Thus the

---

[4]The second duplication was noted in the remarks after Jones' table, but not referred to correctly in its last column.

Table 7.3 The weights of the traces contributing to the λ-coefficients of X for a 4-braid

	$\widetilde{W}_{1111} = \frac{1}{[2][3][4]} \times$	$\widetilde{W}_{211} = \frac{q}{[2][1][4]} \times$	$\widetilde{W}_{22} = \frac{q^2}{[2][2][3]} \times$	$\widetilde{W}_{31} = \frac{q^3}{[2][1][4]} \times$	$\widetilde{W}_4 = \frac{q^6}{[2][3][4]} \times$
1	1	1	1	1	1
$-\lambda$	$q^2 + q^3 + q^4$	$1 + q^2 + q^3$	$1 + q + q^2$	$1 + 1/q + q^2$	$1 + 1/q + 1/q^2$
λ^2	$q^5 + q^6 + q^7$	$q^3 + q^2 + q^5$	$q + q^2 + q^3$	$1/q + q + q^2$	$1/q + 1/q^2 + 1/q^3$
$-\lambda^3$	q^9	q^5	q^3	q	$1/q^3$

Each table entry must be multiplied with the factor on the right in the first row to obtain the contribution of the corresponding trace to $-\sqrt{\lambda}^{e-3}$ times the power of $-\lambda$ in the first column. In the first row the symbol $[i]$ denotes $1 - q^i$

relations between the X_i will hold whenever $MFW \leq 4$, and are useless as a 4-braid test. In a similar vein, one has

Proposition 7.3 *If a braid β has $MFW(\beta) \leq 4$, then $V(\hat{\beta})$ and $\Delta(\hat{\beta})$ together with the exponent sum $[\beta] = e$, determine $P(\hat{\beta})$.*

Proof From e, V and Δ we know

$$\tilde{X}_a = \sum_{i=0}^{3} X_i(q)\, q^{a(i+(e-3)/2)},$$

for $a \in \{-2, -1, 0, 1\}$. (For $a = -2$ and $a = 0$ we have the trivializing substitutions, $a = -1$ corresponds to Δ and $a = 1$ to V.) So one can recover X_i from $\tilde{X}_a$ and e. (One can write down an explicit formula easily.)   □

This condition is thus equally unhelpful as a 4-braid test. For a similar reason, I expect (though I have not rigorously derived) an explanation of the (experimentally observed) failure of Jones' conditions [J, §8] (see formula (8.10)) to obstruct to a 4-braid.

Example 7.3 The knot 11_{386}, known from [LM], has the Jones polynomial of the figure-8-knot. So 11_{386} and its mirror image show that the dependence of P on e in Proposition 7.3 is essential.

Question 7.2 Are there 5-braids $\beta_{1,2}$ (at least one of which has $MFW = 5$), with the same exponent sum, $V(\hat{\beta})$ and $\Delta(\hat{\beta})$, but different $P(\hat{\beta})$?

The lack of such examples, after some check in the knot tables, is at least not fully explainable. Only 6-braids could be found.

Example 7.4 The knots 16_{443392} and $!15_{223693}$ have the same V and Δ but different P polynomial. They have $MFW = 5$ resp. 4, with 5- resp. 4-braid representations of exponent sum 2 resp. -1, so one obtains 6-braids of exponent sum 1 by stabilization.

One can now, as before, go over to the norms in each row in Table 7.3, or of arbitrary linear combinations of such rows. Then for q where J_0 is definite we obtain again estimates on $|X_i(q)|$, or of $|X(q, \lambda)|$ for any non-zero complex number λ. (In particular, for $\lambda = q$ we obtain Theorem 7.3 and for $\lambda = 1/q$ Corollary 7.3.) Although they still contain the Jones and Alexander polynomial conditions (which were observed both non-trivial in comparison to MFW), such skein polynomial norm estimates have not proved in practice, as a 4-braid test, an efficient improvement over their special cases and MFW. Below we will explain how to do much better.

Knots like 10_{132} and 10_{156} in Example 7.2 show a disadvantage of our test, resulting from not taking into account information of other invariants. On the opposite hand, the mere use of Δ or P reduces calculation complexity, and excludes any potential further knot with such polynomial. In the case of P this gives yet a different way to answer negatively Birman's question if one can realize any skein polynomial by a link making MFW sharp (see [St3, St2]).

Using the Brandt–Lickorish–Millett-Ho polynomial, Murakami [Mr] and later Kanenobu [K] gave with Theorem 4.7 a more efficient (in excluding examples, though less in calculation complexity) test for a 3-braid. But the work in [St2] and in the previous chapters of this monograph makes the study of polynomials of 3-braids anyway less relevant. 4-braids become much more of interest, and in this case, after MFW, the problem to find applicable conditions has been largely unsettled for quite a while.

7.5   Recovering the Burau Trace

7.5.1   Conditions on the Eigenvalues

Using just norms clearly weakens the conditions considerably, and so one would like to identify the Burau eigenvalues directly. However, the relations between the X_i do not allow to recover by simple algebraic means the individual traces from P.

Using Squier's unitarity, there is an analytic way to recover the Burau trace of 4-braids (at least for generic t and up to finite indeterminacy). Since by the previous remarks the use of P is not essential, we will describe the procedure given Δ, V and e. This gives the most significant practical enhancement to the above 4-braid tests.

In the following we fix a 4-braid β of exponent sum e, whose closure link has Jones polynomial V and Alexander polynomial Δ. We let t be a unit norm complex number with non-negative real part.

We use that

$$\operatorname{tr} \psi_3 \wedge \psi_3 = (-t)^{e/2} \overline{\operatorname{tr} \psi_3} (-t)^{-e/2}.$$

Then we have

$$\Delta(t) = \frac{1-t}{1-t^4} \left(-\frac{1}{\sqrt{t}}\right)^{e-3} \cdot \left[1 - \operatorname{tr} \psi_3 + (-t)^{e/2} \overline{\operatorname{tr} \psi_3} (-t)^{-e/2} - (-t)^e\right].$$

So, with $i = \sqrt{-1}$,

$$\delta := -\frac{1}{2i}(-t)^{-e/2} \left[\Delta(t)\frac{1-t^4}{1-t}\left(-\frac{1}{\sqrt{t}}\right)^{3-e} - 1 + (-t)^e\right]$$

is a real number. Now when $\lambda_{1,2,3}$ are the eigenvalues of $\psi_3(\beta)$, we have

$$A = \lambda_1 + \lambda_2 + \lambda_3 = (-t)^{e/2}(y + i\delta)$$

$$B = \lambda_1\lambda_2 + \lambda_1\lambda_3 + \lambda_2\lambda_3 = (-t)^{e/2}(y - i\delta)$$

$$C = \lambda_1\lambda_2\lambda_3 = (-t)^e.$$

Here y is a real number we do not know, and we try to determine. Since $|\lambda_k| = 1$, we have as before $|A| \leq 3$ and $|B| \leq 3$. So the range for y is $[-y_0, y_0]$ with $y_0 = \sqrt{9 - \delta^2}$. (If $|\delta| > 3$ we are done as before.) Then

$$\operatorname{tr} \psi_3 \in \left[(-t)^{e/2}(-y_0 + i\delta),\ (-t)^{e/2}(y_0 + i\delta) \right],$$

with the interval understood as a line segment lying in $\mathbb{C}$. Let ψ_- and ψ_+ be the endpoints of this interval.

Now one can restrict the interval $[-y_0, y_0]$ for y using the Jones polynomial as follows.

Let $\rho := (-t)^{-e/2}\operatorname{tr}\bar\psi_2 \in [-2, 2]$. The restriction to the given range follows because $J^{[3]}$ (see the proof of Proposition 7.1) is also definite when $J^{[4]}$ is. From (7.13) we have

$$\operatorname{tr}\psi_3 = \frac{1-t^2}{t(1-t^3)} \left[V(t)\left(-\frac{1}{\sqrt{t}}\right)^{e-3} - \frac{1-t^5}{1-t^2} - \frac{t^2}{1+t}(-t)^{e/2} \cdot \rho \right] =: \tilde\psi(\rho),$$

so

$$\tilde\psi(\rho) = \frac{1-t^2}{t(1-t^3)} \left[V(t)\left(\frac{-1}{\sqrt{t}}\right)^{e-3} - \frac{1-t^5}{1-t^2} \right] - \frac{\rho}{1+2\Re t}(-t)^{e/2}.$$

(Just for the purpose of defining $\tilde\psi$, we should regard here ρ as a formal parameter, rather than as a concrete value.)

Let $\tilde\psi(\pm 2) =: \tilde\psi_\pm$. Since $1 + 2\Re t > 0$, we have $\Re((-t)^{-e/2}\tilde\psi_+) > \Re((-t)^{-e/2}\tilde\psi_-)$.

Then $[\tilde\psi_-, \tilde\psi_+] \subset \mathbb{C}$ is an interval of the same slope as $[\psi_-, \psi_+]$, so we check if they overlap.

Let

$$\tilde y_\pm = (-t)^{-e/2}\tilde\psi_\pm - i\delta.$$

Then for a consistent restriction on $\operatorname{tr}\psi_3$ the following holds:

1. $\tilde y_\pm$ are real
2. $\tilde y_+ \tilde y_- \leq 0$ or $\min(\tilde y_\pm^2 + \delta^2) \leq 9$.

Potentially these conditions may be violated, but in practice they seem always to hold. (We have not elaborated on why this is so, though it may be worth understanding.) Then at least, we consider

$$y \in [-y_0, y_0] \cap [\tilde y_-, \tilde y_+]. \tag{7.17}$$

We have now the cubic

$$x^3 + ax^2 + bx + c := x^3 - Ax^2 + Bx - C = 0.$$

One solution is obtained by Cardano's formula, in MATHEMATICA™

$$\lambda_1 = -\frac{a}{3} - \frac{\sqrt[3]{2}(-a^2 + 3b)}{3\Gamma} + \frac{\Gamma}{3\sqrt[3]{2}},$$ (7.18)

where

$$\Gamma = \sqrt[3]{-2a^3 + 9ab - 27c + \sqrt{27\left(-a^2b^2 + 4b^3 + 4a^3c - 18abc + 27c^2\right)}}.$$ (7.19)

(Note that (7.18) does not depend on which *square* root is taken in (7.19). Also, every solution of the cubic is an eigenvalue, thus which *cube* root is taken in (7.19) does not matter either.)

We must have that $|\lambda_1| = 1$. Then we must check $|\lambda_{2,3}| = 1$. For this their exact determination is not necessary. We have $\lambda_2 + \lambda_3 = A - \lambda_1$ and $\lambda_2\lambda_3 = C/\lambda_1$. In order $|\lambda_{2,3}| = 1$, we must have

$$\lambda_2 + \lambda_3 = c \cdot \sqrt{\lambda_2\lambda_3},$$

with $c \in [-2, 2]$, so $A - \lambda_1 = c\sqrt{C/\lambda_1}$, which is equivalent to

$$\frac{(A - \lambda_1)^2\lambda_1}{C} \in [0, 4].$$ (7.20)

7.5.2   *Applications and Examples*

The image of the r.h.s. of (7.17) under (7.18) will be some curve in $\mathbb{C}$ that generically intersects S^1 in a finite number of points. This parametric equality can be examined numerically and allows to recover $A = \mathrm{tr}\,\psi_3$ up to finite indeterminacy. In particular, we have

Proposition 7.4 *Assume for some $t \in S^1$ with $|\arg t| \leq \pi/2$ we cannot find y as in (7.17), such that λ_1 given by (7.18) has norm 1 and (7.20) holds. Then there is no $\beta \in B_4$ with the given e whose closure has the given Alexander and Jones polynomial.*   □

Remark 7.5 To apply the test in practice, one chooses a small stepwidth s for y in the interval (7.17), and calculates the derivative of the r.h.s. of (7.18) in y to have an error bound on $||\lambda_1|-1|$ in terms of s. When some of the radicands in (7.19) becomes close to 0, some care is needed. One situation where such degeneracy occurs are the knots 15_{144634}, 15_{144635}, and 15_{145731}. They have representations $\beta \in B_4$, whose Burau matrix is trivial for $t = e^{\pi i/3}$. This may be noteworthy on its own in relation to the problem whether ψ_3 is faithful. (Clearly ψ_3 is unfaithful at any root of unity on the center of B_4, but here β is not even a pure braid.)

Example 7.5 Applying Proposition 7.4 we can exclude 11_{387}, one of the 7 prime 11 crossing knots with $b = 5$ but $MFW \leq 4$. Eleven prime knots with 12 crossings, and 63 of 13 crossings where braid index 4 is not prohibited by MFW can be ruled out. The correctness of these examples was later verified by the 2-cabled MFW. Up to 16 crossings more than 4000 examples were obtained. (Let us note that from them only about 100 can be identified using the norm estimates.)

Example 7.6 The check of prime 14 crossing knots to which our criterion applied, revealed six knots with 2-cable MFW bound 8. One, 14_{22691}, can be excluded from braid index 4 (as done with 14_{45759} in [St3]), by making the v-degrees of the polynomial of the 2-cable contradict the exponent sum of its possible 8-braid representation. However, for the other five knots, 14_{28220}, 14_{30960}, 14_{41334}, 14_{41703}, and 14_{44371}, the argument fails, and so our condition seems the only applicable one. (Clearly a 3-cable polynomial is not a computationally reasonable option, and even the 2-cable requires up to several hours, while our test lasts a few seconds.)

Still our criterion leaves open several interesting examples of (apparent) failure of the 2-cable MFW inequality, among them the knot 13_{9684} encountered with M. Hirasawa. A more general, and important, possible application is as follows.

Remark 7.6 In relation to Jones' conjecture 7.1, we already quoted (in the proof of Theorem 7.1) the observation in [St3] that counterexamples to the conjecture would make MFW and *all* of its cabled versions unsharp. In that sense, our 4-braid test may be the first possible approach toward identifying such a counterexample. Birman-Menasco claimed indeed a family of 6-string potential counterexamples, and K. Kawamuro gave later a simpler family on five strings. Extensive checks with our test of Kawamuro's knots failed to turn up successful cases. This puzzled me a while, until Kawamuro reported recently that in fact H. Matsuda falsified all Birman-Menasco (and hence also Kawamuro's) candidates.

7.6 Mahler Measures

7.6.1 3-Braids

The material in Chapter 7 originated from a question of S. Kamada whether the Alexander polynomial Mahler measure $M(\Delta)$ is bounded on closed 3-braids.

Definition 7.1 For a Laurent polynomial $p \in \mathbb{Z}[t^{\pm 1}]$, and $n \in \mathbb{N}_+$, we define the n-norm of p by

$$\|p\|_n := \sqrt[n]{\sum_{k=-\infty}^{\infty} |[p]_k|^n},$$

and its Mahler measure by

$$M(p) := \prod_{|t| \geq 1, p(t) = 0} |t| .$$

This is extended to $p \in \mathbb{Z}[t^{\pm 1/2}]$ in the obvious way. See [SSW, CK].

Kamada's question is related to controlling $|V(t)|$ and $|\Delta(t)|$ for $t \in S^1$, since the 2-norm $\| . \|_2$ and the Mahler measure M of polynomials have circle integral formulas (see (7.21) for the norm). One can thus ask whether $\|\Delta \cdot W_n\|_2$ is bounded for proper $W_n \in \mathbb{Z}[t, 1/t]$, or, (weaker) whether $M(\Delta)$ is bounded (and similarly $M(V)$ and $\|V \cdot W_n\|_2$) for braid index $\leq n$.

For values of t where J_0 is not definite, however, there seems little one can say on the range (on closed 3-braids) of $|V(t)|$ or $|\Delta(t)|$. Likely they are dense in $\mathbb{R}_+$. We can conclude boundedness properties in special cases where the indefinite J_0 values of t are controllable. For example, the following can be proved easily. (Note that for polynomials with integer coefficients the properties a set of polynomials to have bounded 2-norms, or finitely many distinct 2-norms, are equivalent, and also equivalent to the same two properties for the 1-norm.)

Proposition 7.5 *The set*

$$\{ \|(1 - t^{2n})\Delta_{\hat{\beta}}(t)\|_2 \ : \ \beta \in B_{2n}, \ \Delta_{\hat{\beta}} \in \mathbb{Z}[t^{\pm n}] \}$$

is finite for any $n \geq 2$.

Proof We have

$$\|X\|_2^2 = \int_0^1 \left| X(e^{2\pi i s}) \right|^2 ds . \tag{7.21}$$

If $\Delta \in \mathbb{Z}[t^{\pm n}]$, then this integral for

$$X = \frac{(1 - t^{2n})\Delta(t)}{1 - t} = \det(1 - \psi_{2n-1})$$

is controlled from its part over $0 < s < \dfrac{1}{2n}$, and this in turn is controlled by the unitarity of ψ_{2n-1}. To eliminate the extra factor $Q = \dfrac{1 - t^{2n}}{1 - t}$, observe that the norm in (7.21) is just the one in $L^2(S^1)$, and so $\|X/Q\|_2 \leq \|X\|_2 \cdot \|Q^{-1}\|_2$, while $1/Q$ is quadratically integrable over S^1.   □

However, in general a bound on the Mahler measure of arbitrary polynomials of closed n-braids does not exist even for $n = 3$.

Proposition 7.6 *Let $\beta \in B_3$ with $[\beta] = 0$ and $\hat{\beta}$ a knot. Then $M(V(\widehat{\beta^n})) \to \infty$ and $M(\Delta(\widehat{\beta^n})) \to \infty$.*

Thus the Mahler measure of Δ and V is unbounded already for amphicheiral 3-braid knots, e.g., the closures of $(\sigma_1\sigma_2^{-1})^n$ (for $3 \nmid n$; see also Remark 7.7 below). The condition on β was chosen to be fairly general, although the statement can be further extended.

Proof For $[\beta] = 0$, the formula (7.15) becomes with $\psi = \psi_2$,

$$V(\hat{\beta}) = t + \frac{1}{t} + \operatorname{tr}\psi(\beta). \tag{7.22}$$

Because of the row-column symmetry of $\boxplus$, we have (see Section 7.2)

$$\psi(\beta)(t) = (-t)^e \psi(\beta^{-1})(t).$$

By looking at (7.22) and using that $e = [\beta] = 0$ and $V(\hat{\beta}^{-1})(t) = V(\hat{\beta})(t^{-1})$, we see that $\operatorname{tr}\psi(\beta) \in \mathbb{Z}[t, t^{-1}]$ is a self-conjugate polynomial. Thus

$$\operatorname{tr}\psi(\beta)(t) = P(\alpha) \in \mathbb{Z}[\alpha] \quad \text{for} \quad \alpha = \alpha(t) = t + \frac{1}{t}. \tag{7.23}$$

The polynomial P is not constant, since otherwise $V(\hat{\beta}) = t + \frac{1}{t} + p_0$ for some $p_0 \in \mathbb{Z}$, but (e.g., from [J, §12]) no knot has such Jones polynomial.

Let now $\lambda_1, \lambda_2(\beta)$ be the two eigenvalues (precisely speaking, Jordan form diagonal entries) of $\psi(\beta)$. We have $\lambda_1 + \lambda_2 = P(\alpha)$ and $\lambda_1 \cdot \lambda_2 = 1$ (since $[\beta] = 0$), thus

$$\lambda_{1,2}(\alpha) = \frac{P(\alpha)}{2} \pm \sqrt{\frac{P(\alpha)^2}{4} - 1}.$$

(For $P(\alpha) \neq \pm 2$ these are distinct, and $\psi(\beta)$ is indeed diagnoalizable.) Then for $L_n = \widehat{\beta^n}$ we have

$$V_n := V(L_n) = t + \frac{1}{t} + \lambda_1^n + \lambda_2^n = \alpha + \lambda_1(\alpha)^n + \lambda_2(\alpha)^n. \tag{7.24}$$

Let

$$\lambda_1(\alpha) = \lambda(\alpha), \quad \lambda_2(\alpha) = \lambda(\alpha)^{-1}.$$

Next we need an $\alpha \in \mathbb{R} \setminus [-2, 2]$ with $|P(\alpha)| \le 2$. By putting $t = 1$ (and $\alpha = 2$) in (7.22) and using $V(1) = 1$ for a knot, we have

$$1 = 2 + P(\alpha), \quad \text{and so} \quad P(2) = -1.$$

Thus by continuity there exist infinitely many $t \in \mathbb{R}$ with $|t| > 1$ such that $t + \frac{1}{t} = \alpha \in \mathbb{R} \setminus [-2, 2]$ has $|P(\alpha)| \le 2$.

Now note that when $|P(\alpha)| \leq 2$ (and $P(\alpha) \in \mathbb{R}$) then $|\lambda_i(\alpha)| = 1$. Note also that the condition

$$\lambda'(\alpha) = \frac{d\lambda}{d\alpha}(\alpha) = 0$$

implies that α is the zero of some polynomial. Hence there are only finitely many α satisfying it, and we may w.l.o.g. choose (infinitely many) t so that $\lambda'(\alpha(t)) \neq 0$.

Therefore, we have many t with

$$|t| > 1 \text{ such that } |\lambda_i| = 1 \text{ and } \lambda'(\alpha(t)) \neq 0. \tag{7.25}$$

We claim now the following lemma.

Lemma 7.1 *Let α have $|\lambda(\alpha)| = 1$ and $\lambda'(\alpha) \neq 0$. Then there exist zeros α_i of*

$$V_i(\alpha) = \alpha + \lambda(\alpha)^n + \lambda(\alpha)^{-n},$$

i.e., $V_i(\alpha_i) = 0$, such that $\alpha_i \to \alpha$.

This lemma would prove Proposition 7.6 for V thus. By (7.25) and for a small (but fixed) $\varepsilon > 0$, for any natural number l, one can fix l pairwise distinct numbers $t_1, \ldots, t_l \in \mathbb{C}$ with

$$|t_i| > 1 + \varepsilon \quad \text{and} \quad |\lambda(\alpha(t_i))| = 1.$$

Then by the lemma one can find a sequence $\big((z_{n,1}, \ldots, z_{n,l})\big)_n$ of l-tuples of zeros $z_{n,i}$ of V_n (necessarily distinct when n is large enough) with $(z_{n,1}, \ldots, z_{n,l}) \to (t_1, \ldots, t_l)$. Therefore, $M(V_n) \geq (1 + \varepsilon)^l$ when n is large enough.

Proof of Lemma 7.1 By contradiction. The function $\alpha \mapsto \lambda(\alpha)$ is non-constant holomorphic, thus locally invertible for $\lambda'(\alpha) \neq 0$, and (pre)images of open sets are open.

Let thus $\eta = \lambda(\alpha)$ with

$$\eta \in S^1 = \{z \in \mathbb{C}, |z| = 1\}.$$

Choose a bijective restriction

$$\lambda : \overline{O} \to \overline{\mathcal{K}}$$

from the closure of an open neighborhood $O \ni \alpha$ to the closed ε-ball $\mathcal{K} := B(\eta, \varepsilon)$ around η. Assume by contradiction that

$$\lambda^{-1}(\mathcal{K}) \cap \{\text{zeros of } V_{n_i}\} = \varnothing \tag{7.26}$$

for a subsequence (V_{n_i}) of (V_n).

By perturbing η and adjusting ε we may w.l.o.g. assume that (with $i = \sqrt{-1}$)

$$\mathcal{K} = \overline{B(\eta, \varepsilon)} \not\ni \pm i, 0. \tag{7.27}$$

Thus

$$L = -\overline{O} = \{-\alpha : \lambda(\alpha) \in \overline{\mathcal{K}}\}$$

does not contain 0, and L is a closed set. Moreover, since $|\alpha| \to \infty$ gives $\lambda(\alpha) \to 0$, we see that L is bounded, and thus compact.

Let $\mathcal{K}_0 := B\left(\eta, \dfrac{\varepsilon}{2}\right)$. Then because of (7.27) we have

$$0 \notin \mathcal{K}_0. \tag{7.28}$$

We see first that for n sufficiently large,

$$\{\lambda(\alpha)^n + \lambda(\alpha)^{-n} : \alpha \in \lambda^{-1}(\mathcal{K}_0)\} \supset L. \tag{7.29}$$

To do so, notice that since $|\eta| = 1$, for n large enough, $\{\lambda(\alpha)^n : \lambda(\alpha) \in \mathcal{K}_0\} = \mathcal{K}_0^n$ contains every bounded set $\mathcal{M}$ with $\mathrm{dist}(\mathcal{M}, 0) > 0$. Let this set be then

$$\mathcal{M} := \left\{\delta \in \mathbb{C} \setminus \{0\} : \delta + \frac{1}{\delta} \in L\right\},$$

which is compact because L is compact. Moreover, $0 \notin \mathcal{M}$, and two disjoint compact sets have positive distance.

We claim now that when (7.29), then the polynomial (7.24), which we rewrite as

$$V_n = \alpha + \lambda(\alpha)^n + \lambda(\alpha)^{-n},$$

has a zero α_n with $\lambda(\alpha_n) \in \overline{\mathcal{K}_0} \subset \mathcal{K}$ for all sufficiently large n, in contradiction to (7.26).

To see the claim, fix any n with (7.29). Let $\eta_1 = \lambda(\alpha_{n,1}) \in \mathcal{K}_0$ be chosen in some (arbitrary) way. Thus $-\alpha_{n,1} \in L$. By (7.29), there is an $\alpha_{n,2} \in \lambda^{-1}(\mathcal{K}_0)$ with

$$\lambda(\alpha_{n,2})^n + \lambda(\alpha_{n,2})^{-n} = -\alpha_{n,1}.$$

Next, since $-\alpha_{n,2} \in L$, there is an $\alpha_{n,3} \in \lambda^{-1}(\mathcal{K}_0)$ with

$$\lambda(\alpha_{n,3})^n + \lambda(\alpha_{n,3})^{-n} = -\alpha_{n,2},$$

etc. Now, the property (7.28) we have chosen for $\mathcal{K}_0$ implies that $\lambda^{-1}(\overline{\mathcal{K}_0})$ is compact. Thus $(\alpha_{n,j})_j$ has a limit point $\alpha_n = \lim_{k \to \infty} \alpha_{n,j_k}$ inside. This point α_n then satisfies

$$\lambda(\alpha_n)^n + \lambda(\alpha_n)^{-n} = -\alpha_n,$$

as desired.   □

This shows the case of V. For Δ the argument is similar but simpler. The formula is

$$\Delta_n := \Delta(L_n) = \frac{1}{\alpha + 1}(2 - \lambda^n(\alpha) - \lambda^{-n}(\alpha)).$$

Thus we look at the $\alpha \neq -1$ for which $\lambda(\alpha)$ are roots of unity. Since $\lambda : \alpha \mapsto \lambda(\alpha)$ is a continuous map, we see that the roots of Δ are dense in the set of t for which α from (7.23) lies in $\lambda^{-1}(S^1) \not\subset [-2, 2]$.   □

Remark 7.7 Note from (7.22) and (7.23) that all links $\hat{\beta}$ for $\beta \in B_3$ with $[\beta] = 0$ have self-conjugate Jones (and skein) polynomial. Many are not amphicheiral, thus giving examples where these polynomials fail to detect chirality. The simplest such knot is 10_{48}, the closure of $[1^3 - 2^4 1^2 - 2]$.

Remark 7.8 Dan Silver pointed out that for the Alexander polynomial a proof can be obtained (possibly more elegantly, but not immediately) from the ideas in [SW].

7.6.2 Skein Polynomial and Generalized Twisting

By using the representation theory, one can extend the scope of the results in [SSW, CK] on bounded Mahler measure to (parallel) multi-strand twisting. We give a version for the skein polynomial, since its special case, the Jones polynomial, was discussed in [CK]. One can obtain this special case by setting $\lambda = q$ in the below theorem. (The reversal of component orientation is not a serious problem for V, unlike for P.) We write as before $P_L(v, z)$ for the skein polynomial and use the skein rule $v^{-1}P_+ - vP_- = zP_0$.

A *template* T consists, as in [ST], of a number of strands, that create crossings and *slots*, only that for us strands are oriented and slots have the more general form (with an arbitrary number of in- and outputs)

 .

A(n oriented) diagram D is *associated* to T, if D is obtained from T by inserting into each slot of T a braid of a certain number of full twists (on the proper number of strands).

The 1-norm $||P||_1$ of a (Laurent) polynomial P (even in several variables) is understood to be the sum of absolute values of all its (non-zero) coefficients (taken in all variables). We write $M(P)$ for its Mahler measure. It is known that for polynomials P (of real coefficients), $M(P) \leq ||P||_1$ (see §2 of [SSW]).

Theorem 7.5 *For each n there is a polynomial $D_n(q)$ such that for each template of $n_1, \ldots, n_k$-strand parallel twist slots and each diagram D associated to T we have*

$$\left\| \prod_{i=1}^{k} D_{n_i}(q) \ \cdot \ P_D(\lambda, \sqrt{q} - 1/\sqrt{q}) \right\|_1 \leq C_T,$$

where C_T is a constant that depends only on T. Furthermore D_n are made of a product of terms $1 - q^l$ for some $1 \leq l \leq n$. In particular,

$$\{ M(P_D(\lambda, \sqrt{q} - 1/\sqrt{q})) \ : \ D \text{ is associated to } T \}$$

is bounded.

Proof Write $D(t_1, \ldots, t_k) = D_{t_1, \ldots, t_k}$ for the diagram associated to T by putting into the ith slot t_i full twists (on n_i strands).

Let first $k = 1$. Then the complement of the slot is a room with $n = n_1$ parallel in- and outputs. The skein module of such a room is generated by (positive permutation) braids, and thus it suffices to work with the n-strand braid group. Again, by [J, Proposition 6.2] (we resume references to Jones' paper)

$$P_L(\lambda, \sqrt{q} - 1/\sqrt{q}) = X_L(q, \lambda^2/q),$$

where X_L is a writhe-strand normalized (Definition 6.1 in [J]) weighted trace sum (Equation (5.5)), and the full twist has the effect of multiplying the traces by a power of q (Lemma 9.3).

All terms in the denominator of the (rational) generating series

$$\sum_{t=0}^{\infty} X(D_t)(q, \lambda^2/q) \, x^t \tag{7.30}$$

are obtained from denominators in the definition of $X_L(q, \lambda)$ and quantities that differ by the number $t = t_1$ of twists. To identify these terms we will work up to units in $\mathbb{Z}[q^{\pm 1/2}, \lambda^{\pm 1/2}]$ that do not depend on the number t of twists but just on T. The quantities depending on t are only the power of q in the traces and the term $\sqrt{\lambda}^e$ in Definition 6.1, where e is the exponent sum. Since the full twist has $e = n(n-1)$, we find the terms $1 - \lambda^{n(n-1)/2} q^j x$ for some values of $j \geq 0$. The exponent $n(n-1)/2$ becomes $n(n-1)$ and $j \geq -n(n-1)/2$ when replacing λ^2/q

for λ. It suffices now to collect the denominator terms occurring in the definition of X_L (and replace λ^2/q for λ). From Definition 6.1 we have $\lambda^{(n-1)/2}$ (which is a unit, fixed for given T, and remains so when replacing λ^2/q for λ) and some power of $1 - q$. From the text below Figure 5.6 and (5.5) we have some power of $1 - \lambda q$ (that becomes $1 - \lambda^2$ under substitution) and (in $Q(q)$) products of terms $1 - q^l$ for some $1 \leq l \leq n$ (with l being a hook length of a box in a Young diagram of n boxes). Thus the denominator of (7.30) is something we can take to be $D_n(q)$ times a product of terms of the form $1 - \lambda^2$ and $1 - \lambda^{n(n-1)}q^l x$.

Now if $k > 1$, we first observe that $X(D_{t_1,\dots,t_k})(q,\lambda)$ satisfy a linear recurrence (with coefficients in $\mathbb{Z}[q^{\pm1/2}, \lambda^{\pm1/2}]$) in t_i for any fixed value of $t_j, j \neq i$. Moreover that recurrence itself does not depend on $t_j, j \neq i$ for fixed i, only its initial values do depend. Then one inductively argues over k that the generating series

$$\sum_{t_1,\dots,t_k \geq 0} X(D_{t_1,\dots,t_k})(q,\lambda^2/q) \; x_1^{t_1} \dots x_k^{t_k}$$

has a denominator which is the product of the D_{n_i} and terms $1 - \lambda^2$ and $1 - \lambda^{n_i(n_i-1)}q^j x_i$ for $1 \leq i \leq k$ and $j \geq -n_i(n_i - 1)/2$. The cases when $t_i < 0$ are analogous. From this one easily concludes the claim up to the factors $1 - \lambda^2$. Since for fixed T the power p_T of $1 - \lambda^2$ is fixed, and the λ-span of P_D is bounded by the Morton–Franks–Williams inequality (see Proposition 15.1 of [J]), one can get disposed of the $1 - \lambda^2$ factors by linear combinations of the coefficients of $(1 - \lambda^2)^{p_T} P_D \cdot \prod D_{n_i}$ in the powers of λ.   □

7.7   Knots with Unsharp MFW Inequality

This brief part was motivated by the paper of Kawamuro [Kw], which I came across during the continuous work on my monograph. Kawamuro was interested in finding infinitely many knots for which the MFW inequality is strict, in particular such satisfying Jones' conjecture in Section 7.1. The point to make is that we have at least two ways allowing us to obtain her, and further such knots much more easily. (In particular it is not necessary to appeal to the heavy geometric machinery of Neumann–Rudolph–Giroux.)

We already saw an infinite family of examples with unsharp MFW (and satisfying Jones' conjecture) in the proof of Theorem 7.1. This family is, of course, little insightful, since it is settled already by Birman–Menasco's 3-braid work. The 4-braid examples they proposed, but could not decide about, deserve more attention. As given in Figure 4 of [Kw], consider for $(x, y, z, w) \in \mathbb{Z}^4$ the 4-braids $\beta_{x,y,z,w} = [212^2 13^x 2^y - 12^z 3^w]$, and let $K_{x,y,z,w} = \hat{\beta}_{x,y,z,w}$. Birman-Menasco observe that $MFW(K_{x,y,z,w}) \leq 3$ (see Lemma 2.9 In [Kw]), but suspect that many of the $K_{x,y,z,w}$ have braid index 4. One can exhibit an infinite family of $K_{x,y,z,w}$ of $b = 4$, and thus recover Theorem 2.8 of [Kw], like this.

Proposition 7.7 *There is a natural number N and six 4-tuples $(\alpha_1, \ldots, \alpha_4) \in \mathbb{Z}^4$ of different mod-2-reductions in $(\mathbb{Z}/2)^4$, such that if $K_\gamma := K_{\gamma_1, \ldots, \gamma_4}$ is a knot, then $b(K_\gamma) = 4$ whenever $(\alpha_i) \equiv (\gamma_i)$ mod 2 and $\gcd\left(\dfrac{\gamma_i - \alpha_i}{2}\right) \geq N$.*

Proof Since (complex) roots of unity are dense on the complex unit circle, choose a root of unity t, for which the 3-braid test in Corollary 7.3 applies for 9_{42} or 9_{49} (see Example 7.1). Then, using the formula for Δ in Lemma 3.1 of [SSW], we see that for each such t there is an $n \in 2\mathbb{Z} \setminus \{0\}$, such that $\Delta_{\hat{\beta}}(t)$ is preserved when β is replaced by $\sigma_i^n \beta$.

Moreover, it is easy to check that for all 6 of the 16 possible vectors of parities of (x, y, z, w), for which $K_{x,y,z,w}$ is a knot, there is a representation of (at least one of) 9_{42} or 9_{49} of that parity combination. (The representations given after Definition 2.7 of [Kw] show 3 of the parity types.) Now, take such a representation $\beta_{\alpha_1, \ldots, \alpha_4}$ and vary the parameters α_i by multiples of n. Then, since we can choose (for proper t) n to be any sufficiently large even natural number, we obtain in the claim.   □

Of course, this elementary argument could be further concretified and strengthened. Presumably, one can do much better using J. Murakami's test theorem 4.7 (since it is an equality). Note that our proof does not confirm Jones' conjecture on any $K_{x,y,z,w}$, so it may trigger the question: would it help to find a counterexample? This seems very optimistic, though, as we will soon see from an alternative approach to Kawamuro's theorem, using the work in Section 7.1. First we can settle (also with regard to Jones' conjecture) the concrete examples $K_{-1,-2,m,2}$ for $m \geq 2$ even, obtained in her proof.

Proposition 7.8 *We have $b(K_{-1,-2,m,2}) = 4$ for $m \neq -1, 0$.*

Proof We give just a brief explanation. We try to calculate the skein polynomial of the 2-cable $\beta^{[2]}_{-1,-2,m,2}$ of $\beta_{-1,-2,m,2}$, obtained by replacing σ_i by $\sigma_{2i}\sigma_{2i+1}\sigma_{2i-1}\sigma_{2i}$, and we may consider just the coefficient of $v^{4[\beta_{-1,-2,m,2}]+7}$. Again $\pi_Y([2312])$ has at most seven different eigenvalues δ_i, for all $Y \vdash 8$ taken together. But now we have no full twist, and so do not need to deal with the various Y (and their weights W_Y, etc.) one by one. We can sum $W_Y \cdot c(i, Y)$ over $Y \vdash 8$, and are left with seven different q-rational expressions c_i to determine. Again this can be done from seven explicit polynomials, which can be obtained using Morton's program. (If we focus on one parity of m only, we can again reduce the unknowns c_i and test polynomials to 5, but need to calculate polynomials from slightly more complicated braids.)

We calculated in fact 11 polynomials, for $|m| \leq 5$, and used those 9 for $|m| \leq 4$ in the determination of c_i to have extra safety. The resulting linear equation system for c_i (with matrix $\{\delta_i^m\}$) is now not a serious problem to MATHEMATICA, which gives the solution (for generic q) in just a few seconds. Clearing denominators, and looking already at the extremal q-degrees, we see that for $|m| \geq 6$ among the seven terms $c_i\delta_i^m$ the one for $i = 2$ has the lowest minimal or highest maximal degree (as a unique term, except for $m = 6$, where we must use also that the leading coefficient of $c_6\delta_6^m$ has the same sign). With an explicit check for the other m, we conclude the claim.   □

This method of proof has also the advantage of easily leading to a qualitative improvement of Kawamuro's theorem, which is closer to what one should expect.

Proposition 7.9 *The tuples* (x, y, z, w) *for which* $K_{x,y,z,w}$ *has braid index 4 (and satisfies Jones' conjecture) are generic in* $\mathbb{Z}^4$, *in the sense that*

$$\lim_{n \to \infty} \frac{\left| \{ (x, y, z, w) \in \mathbb{Z}^4 \,:\, b(K_{x,y,z,w}) = 4 \} \cap [-n, n]^4 \right|}{\left| [-n, n]^4 \right|} = 1 . \qquad (7.31)$$

Proof Extending the argument for Proposition 7.8, we see that

$$\tilde{P}(x, y, z, w) := [P(\hat{\beta}^{[2]}_{x,y,z,w})]_{v^{4[\beta x,y,z,w]+7}} (\sqrt{q} - 1/\sqrt{q}) \cdot \sqrt{q}^{\,4[\beta x,y,z,w]+7}$$

$$= \sum_{i_x, i_y, i_z, i_w = 1}^{7} c_{i_x, i_y, i_z, i_w} \delta^x_{i_x} \delta^y_{i_y} \delta^z_{i_z} \delta^w_{i_w} ,$$

where c_{i_x, i_y, i_z, i_w} are again some rational expressions in q. They are obviously not all zero, and the polynomials $\tilde{P}(x, y, z, w)$ have the following property: if one fixes three of the parameters (say x, y, z), and $\tilde{P}(x, y, z, w) = 0$ for seven different values of the remaining parameter (here w), then it vanishes for all values of that parameter (at the given fixed 3 others). It is then not hard to deduce (7.31) from this (see, e.g., the proof of Lemma 11.1 of [St8]).   □

Appendix
Postliminaries

A.1 Fibered Dean Knots (Hirasawa–Murasugi)

Here we present some material due to Hirasawa and Murasugi, who studied fibering of generalized Dean knots. An overview is given in [HM]. As this is work in progress, a longer exposition may appear subsequently elsewhere.

Definition A.1 The Dean knot $K(p, q|r, rs)$ is given by the closed p-braid

$$(\sigma_{p-1}\sigma_{p-2}\cdots\sigma_1)^q(\sigma_1\sigma_2\cdots\sigma_{r-1})^{rs},$$

with $p > r > 1$ and q, s non-zero integers such that $(q, p) = 1$.
Hirasawa and Murasugi proposed a conjecture on these knots and obtained so far the following partial results.

Conjecture A.1 (Hirasawa–Murasugi) A Dean's knot $K(p, q|r, rs)$ is a fibered knot if and only if its Alexander polynomial is monic, that is, $\max \operatorname{cf} \Delta = \pm 1$.

Proposition A.1 (Murasugi) *This conjecture has been proven for the following cases.*

(a) $q = kp + 1$, and r and s are arbitrary,
(b) $q = kp - 1$, and r and s are arbitrary,
(c) $r = p - 1$, and q and s are arbitrary.

The last case implies in particular that the conjecture is true for $p = 3$. Below follows a part of the argument in this case that settles Lemma 3.1. (Other parts of the proof are very similar to some of our previous arguments. It seems, for example, that Hirasawa and Murasugi were to some extent aware of Theorem 3.3.)

Proof of Lemma 3.1 We consider the braids $[(123)^k - 2]$, with $k > 0$ fixed. By isotopies and Hopf plumbings, we modify our surface

© The Author(s) 2017

A. Stoimenow, *Properties of Closed 3-Braids and Braid Representations of Links*,
SpringerBriefs in Mathematics, https://doi.org/10.1007/978-3-319-68149-8

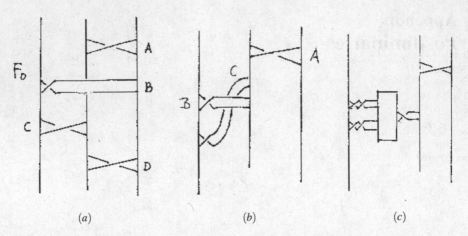

Fig. A.1 Sliding bands I

$$[(123)^k - 2] \to [(1223)^k - 2] = [12(2312)^{k-1}23 - 2] \to$$

$$[(2312)^{k-1}23 - 212] = [(2312)^{k-1}2312 - 1] = [(2312)^k - 1].$$

Let F be the surface of the band representation $\beta = [(2312)^k - 1]$. We will show that F is a fiber surface.

Consider the subsurface F_0 of F that spans $[2312]$ in the natural manner. We deform the k copies of F_0 to the (isotopic sub)surfaces F_0' by a series of diagrams, see Figure A.1. (Here strands are numbered from right to left and words composed downward.)

In Figure A.1, for the move (a) $\to$ (b), we slide B and C, respectively, along D and A, then delete D (by deplumbing a Hopf band), and then slide C back along A. For the move (b) $\to$ (c), we slide C along B, then slide B along C. The k bands A can be subsequently removed by Murasugi desumming a $(2, k)$-torus link fiber surface. Thus the surface F, spanned by $\hat{\beta}$, turns after (de)summing Hopf bands into a surface F' consisting of k copies of F_0' and one negative band N. See Figure A.2a.

In Figure A.2, we perform the move (a) $\to$ (b), sliding B along N. The last surface F', in Figure A.2b, is Murasugi sum of a fiber surface spanning the $(2, 2, -2)$-pretzel link and $\tilde{F}$, where $\tilde{F}$ consists of $k - 1$ copies of F_0' and the band N. By induction on k, we see that F' is a fiber surface, and hence $\hat{\beta}$ is a fibered link.   □

A.2 A-Decomposition (Joint with Hirasawa–Ishiwata)

For the proof of Theorem 1.2 we introduce the A-decomposition due to Kobayashi [Ko].

Fig. A.2 Sliding bands II

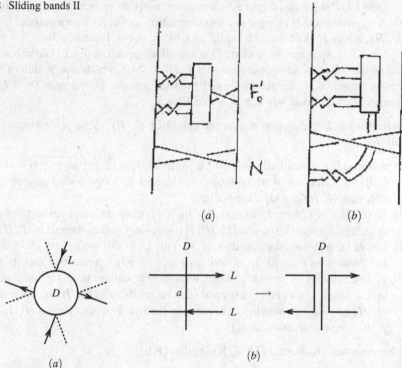

(a)   (b)

(a)   (b)

Fig. A.3 A-operation

A *sutured manifold* in the sense of Gabai [Ga3] can be understood as a pair (L, H) consisting of a closed 3-dimensional submanifold H of $\mathbb{R}^3$ with boundary $S = \partial H$ a connected surface, and a set of oriented loops $L \subset S$, called *suture*. We require that one can orient the connected components of $S \setminus L$ so that the induced orientations on L coincide from both sides of L (in particular a connected component of $S \setminus L$ never bounds to itself along a loop of L), and are given by the orientation of L.

Let F be a connected Seifert surface of a(n oriented) link $L = \partial F$. We embed F as $F \times \{0.5\}$ into the *bicolar* $H = F \times I$ (with $I = [0, 1]$). Then (L, H) becomes a sutured manifold. We call it *canonical sutured manifold* $C(F)$ of F.

We describe some basic operations on sutured manifolds (L, H).

A *decomposition disk* D is a disk with $P = \partial D \subset \partial H$, properly embedded in the complement of H (i.e., $D \cap H = P$). We require that D is not parallel to $S = \partial H$, and satisfies $P \cap F \neq \varnothing$. We assume also that the intersection of P and F is transversal, so that it is a collection of points.

Since $L = \partial F$ is separating on S, the intersection $D \cap L = P \cap F$ is an even number of points, and the orientation of L at the intersection points is alternating (with respect to the orientation of the loop P). See Figure A.3a.

Then $L \cap P$ separates P into a collection of intervals or *arcs*. Let a be such an arc. An *A-operation* on D along a is a transformation of (L, H) into a sutured manifold (L', H), where L' is obtained by splicing L along a. See Figure A.3b.

A *product decomposition* along D is a similar operation, due to Gabai [Ga3], and can be described as an A-operation if $|L \cap D| = 2$ (in which case which of the two arcs is chosen is irrelevant), followed by a subsequent gluing of a $D^2 \times I$ into H along a neighborhood $N(P) \simeq S^1 \times I$ of P on S.

Definition A.2 We define a sutured manifold (L, H) to be *A-decomposable* as follows:

1) Assume H is a standardly embedded handlebody (i.e., so that $\overline{S^3 \setminus H}$ is also one). If L is a collection of trivial loops on ∂H, and all loops bound *disjoint* disks in ∂H, then (L, H) is A-decomposable.
2) If (L', H') is obtained from (L, H) by a product decomposition (along some decomposition disk D), and (L', H') is A-decomposable, then so is (L, H).
3) Let D be a decomposition disk of H with $|L \cap D| = 2n$ and choose among the $2n$ arcs on $P = \partial D$ a collection of n cyclically consecutive arcs $a_1, \ldots, a_n$. (Consecutive is to mean that, taken with their boundary in $L \cap P$, their union is a single interval in P, and not several such intervals.) Let (L_i, H) be obtained from (L, H) by A-decomposition on D along a_i for $i = 1, \ldots, n$. Then if all (L_i, H) are A-decomposable, so is (L, H).

Theorem A.1 (Kobayashi [Ko], Kakimizu [Kk])

1) A fiber surface is a unique incompressible surface.
2) The property a surface to be a unique incompressible surface is invariant under Hopf (de)plumbing.
3) If $C(F)$ is A-decomposable, then F is a unique incompressible surface for $L = \partial F$.

Now we complete the proof of Theorem 1.2.

Proof of Theorem 1.2 Let L be a 3-braid link. If L is split, the splitting sphere also splits any incompressible surface for L. Since L is a split union of a 2-braid link and an unknot (incl. 2 and 3-component unlinks), the claim is easy. Excluding split links, Corollary 4.2 shows we need to consider only connected Seifert surfaces. It also suffices to deal with the non-fibered links only. These are equivalent under Hopf (de)plumbing to some of $(123)^k$ ($k \neq 0$). For such links, we can modify the transformation of Hirasawa and Murasugi from Section A.1. Then we can turn by Hopf (de)plumbing the surfaces into those like in Figure A.2a, consisting of k

Fig. A.4 An
A-decomposable Seifert
surface

copies of F_0', but now *without* the lower band N. (Figure A.4 shows the case $k = 4$.) Then we use *A*-decomposition, as shown in Figures A.5 and A.6. One applies *A*-operations along the arcs $a_{1,2}$. We show only the result for a_1, the case of a_2 and other k is analogous.   □

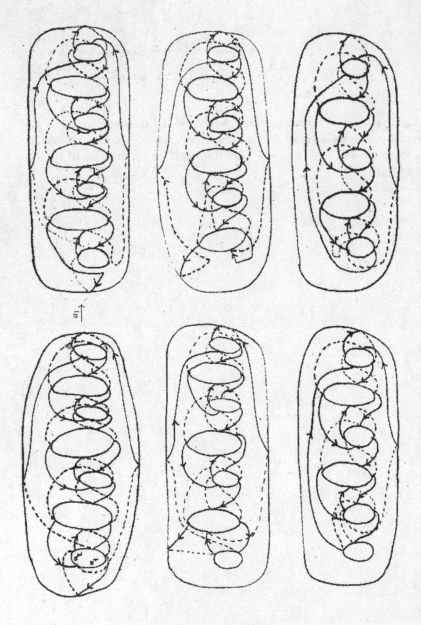

Fig. A.5 This sequence of diagrams describes the A-decomposition along arc a_1 of the canonical sutured manifold corresponding to the surface in Figure A.4. The A-operation is the change between the first and the second diagram. The following diagrams (*left to right in each row*) display isotopies and product decompositions, simplifying the suture. The reduction is continued in Figure A.6. The A-operation along a_2 must be performed, too, but the decomposition is similar

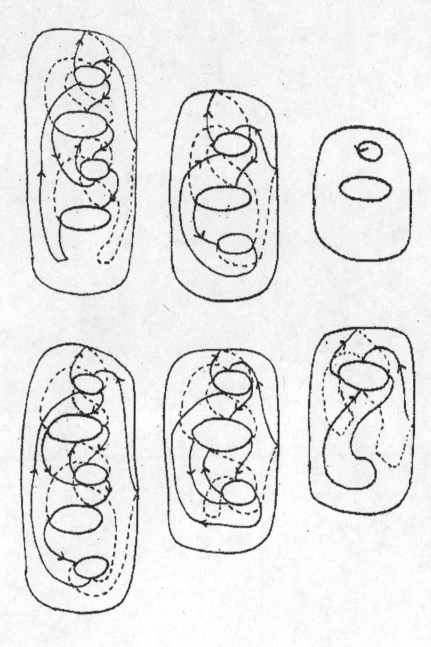

Fig. A.6 The completion of the sequence of suture isotopies and product decompositions of Figure A.5, leading finally to a trivial curve on a torus

References

[Al] J.W. Alexander, A lemma on systems of knotted curves. Proc. Natl. Acad. Sci. USA **9**, 93–95 (1923)

[Al2] J.W. Alexander, Topological invariants of knots and links. Trans. Am. Math. Soc. **30**, 275–306 (1928)

[Ar] E. Artin, Theorie der Zöpfe, Abh. Math. Sem. Hamburgischen Univ. **4**, 47–72 (1926)

[Ar2] E. Artin, Theory of braids. Ann. Math. **48**(2), 101–126 (1947)

[B] J.S. Birman, On the Jones polynomial of closed 3-braids. Invent. Math. **81**(2), 287–294 (1985)

[Be] D. Bennequin, Entrelacements et équations de Pfaff. Soc. Math. de France, Astérisque **107–108**, 87–161 (1983)

[Bi] S. Bigelow, Representations of braid groups, in *Proceedings of the International Congress of Mathematicians*, vol. II (2002), pp. 37–45

[BKL] J.S. Birman, K. Ko, S.J. Lee, A new approach to the word and conjugacy problems in the braid groups. Adv. Math. **139**(2), 322–353 (1998)

[Bl] S.A. Bleiler, Realizing concordant polynomials with prime knots. Pac. J. Math. **100**(2), 249–257 (1982)

[BLM] R.D. Brandt, W.B.R. Lickorish, K. Millett, A polynomial invariant for unoriented knots and links. Invent. Math. **74**, 563–573 (1986)

[BM] J.S. Birman, W.W. Menasco, Studying knots via braids III: classifying knots which are closed 3 braids. Pac. J. Math. **161**, 25–113 (1993)

[BM2] J.S. Birman, W.W. Menasco, Studying links via closed braids II: On a theorem of Bennequin. Topol. Appl. **40**(1), 71–82 (1991)

[BM3] J.S. Birman, W.W. Menasco, Studying knots via braids VI: a non-finiteness theorem. Pac. J. Math. **156**, 265–285 (1992)

[BW] J.S. Birman, H. Wenzl, Braids, link polynomials and a new algebra. Trans. Am. Math. Soc. **313**(1), 249–273 (1989)

[CK] A. Champanerkar, I. Kofman, On the Mahler measure of Jones polynomials under twisting. Algebr. Geom. Topol. **5**, 1–22 (2005)

[Cr] P.R. Cromwell, Homogeneous links. J. Lond. Math. Soc. (series 2) **39**, 535–552 (1989)

[DL] O. Dasbach, X.-S. Lin, On the head and the tail of the colored Jones polynomial. Compos. Math. **142**(5), 1332–1342 (2006)

[EKT] S. Eliahou, L.H. Kauffman, M. Thistlethwaite, Infinite families of links with trivial Jones polynomial. Topology **42**(1), 155–169 (2003)

[Fi] T. Fiedler, A small state sum for knots. Topology **32**(2), 281–294 (1993)

© The Author(s) 2017
A. Stoimenow, *Properties of Closed 3-Braids and Braid Representations of Links*,
SpringerBriefs in Mathematics, https://doi.org/10.1007/978-3-319-68149-8

[Fi2] T. Fiedler, *Gauss Sum Invariants for Knots and Links*. Mathematics and Its Applications, vol. 532 (Kluwer Academic Publishers, Boston, 2001)

[FK] T. Fiedler, V. Kurlin, A one-parameter approach to links in a solid torus. J. Math. Soc. Jpn. **62**(1), 167–211 (2010). math.GT/0606381

[Fu] H. Fujii, Geometric indices and the Alexander polynomial of a knot. Proc. Am. Math. Soc. **124**(9), 2923–2933 (1996)

[FW] J. Franks, R.F. Williams, Braids and the Jones-Conway polynomial. Trans. Am. Math. Soc. **303**, 97–108 (1987)

[F&] P. Freyd, J. Hoste, W.B.R. Lickorish, K. Millett, A. Ocneanu, D. Yetter, A new polynomial invariant of knots and links. Bull. Am. Math. Soc. **12**, 239–246 (1985)

[Ga] D. Gabai, The Murasugi sum is a natural geometric operation, in *Low-Dimensional Topology (San Francisco, California, 1981)*. Contemporary Mathematics, vol. 20 (American Mathematical Society, Providence, RI, 1983), pp. 131–143

[Ga2] D. Gabai, The Murasugi sum is a natural geometric operation II, in *Combinatorial Methods in Topology and Algebraic Geometry (Rochester, N.Y., 1982)*. Contemporary Mathematics, vol. 44 (American Mathematical Society, Providence, RI, 1985), pp. 93–100

[Ga3] D. Gabai, Detecting fibred links in S^3. Comment. Math. Helv. **61**(4), 519–555 (1986)

[Gr] F. Garside, The braid group and other groups. Q. J. Math. Oxford **20**, 235–264 (1969)

[HM] M. Hirasawa, K. Murasugi, Double-torus fibered knots and pre-fiber surfaces. Musubime to Teijigen Topology (Dec. 1999), pp. 43–49

[HT] J. Hoste, M. Thistlethwaite, KnotScape, a knot polynomial calculation and table access program. Available at http://www.math.utk.edu/~morwen

[HTW] J. Hoste, M. Thistlethwaite, J. Weeks, The first 1,701,936 knots. Math. Intell. **20**(4), 33–48 (1998)

[I] M. Ishikawa, On the Thurston-Bennequin invariant of graph divide links. Math. Proc. Camb. Philos. Soc. **139**(3), 487–495 (2005)

[J] V.F.R. Jones, Hecke algebra representations of braid groups and link polynomials. Ann. Math. **126**, 335–388 (1987)

[J2] V.F.R. Jones, A polynomial invariant of knots and links via von Neumann algebras. Bull. Am. Math. Soc. **12**, 103–111 (1985)

[K] T. Kanenobu, Relations between the Jones and Q polynomials of 2-bridge and 3-braid links, Math. Ann. **285**, 115–124 (1989)

[K2] T. Kanenobu, Examples on polynomial invariants of knots and links II. Osaka J. Math. **26**(3), 465–482 (1989)

[K3] T. Kanenobu, Examples on polynomial invariants of knots and links. Math. Ann. **275**, 555–572 (1986)

[K4] T. Kanenobu, An evaluation of the first derivative of the Q polynomial of a link. Kobe J. Math. **5**(2), 179–184 (1988)

[Ka] L.H. Kauffman, An invariant of regular isotopy. Trans. Am. Math. Soc. **318**, 417–471 (1990)

[Ka2] L.H. Kauffman, State models and the Jones polynomial. Topology **26**, 395–407 (1987)

[Ki] M.E. Kidwell, On the degree of the Brandt-Lickorish-Millett-Ho polynomial of a link. Proc. Am. Math. Soc. **100**(4), 755–762 (1987)

[Kk] O. Kakimizu, Classification of the incompressible spanning surfaces for prime knots of ≤ 10 crossings. Hiroshima Math. J. **35**, 47–92 (2005)

[Kn] J.A. Kneissler, Woven braids and their closures. J. Knot Theory Ramifications **8**(2), 201–214 (1999)

[Ko] T. Kobayashi, Uniqueness of minimal genus Seifert surfaces for links. Topol. Appl. **33**(3) , 265–279 (1989)

[Kr] D. Kreimer, *Knots and Feynman Diagrams*. Cambridge Lecture Notes in Physics, vol. 13 (Cambridge University Press, Cambridge, 2000)

[Kw] K. Kawamuro, The algebraic crossing number and the braid index of knots and links. Algebr. Geom. Topol. **6**, 2313–2350 (2006)

[LM] W.B.R. Lickorish, K.C. Millett, A polynomial invariant for oriented links. Topology **26**(1), 107–141 (1987)

[LT] W.B.R. Lickorish, M.B. Thistlethwaite, Some links with non-trivial polynomials and their crossing numbers. Comment. Math. Helv. **63**, 527–539 (1988)

[Mo] H.R. Morton, Seifert circles and knot polynomials. Proc. Camb. Philos. Soc. **99**, 107–109 (1986)

[Mo2] H.R. Morton, (ed.), Problems, in *Braids*, Santa Cruz, 1986, ed. by J.S. Birman, A.L. Libgober. Contemporary Mathematics, vol. 78 (Cambridge University Press, Cambridge, 1986), pp. 557–574

[MP] K. Murasugi, J. Przytycki, The skein polynomial of a planar star product of two links. Math. Proc. Camb. Philos. Soc. **106**(2), 273–276 (1989)

[MP2] K. Murasugi, J. Przytycki, An index of a graph with applications to knot theory. Mem. Am. Math. Soc. **106**(508) (1993)

[Mr] J. Murakami, The Kauffman polynomial of links and representation theory. Osaka J. Math. **24**(4), 745–758 (1987)

[MS] H.R. Morton, H.B. Short, The2-variable polynomial of cable knots. Math. Proc. Camb. Philos. Soc. **101**(2), 267–278 (1987)

[MS2] H.R. Morton, H.B. Short, br9z.p, a Pascal program for calculation of the skein polynomial from braids. http://www.liv.ac.uk/~su14/knotprogs.html

[MT] W.W. Menasco, M.B. Thistlethwaite, The Tait flyping conjecture. Bull. Am. Math. Soc. **25**(2), 403–412 (1991)

[MTr] J. Malesic, P. Traczyk, Seifert circles, braid index and the algebraic crossing number. Topol. Appl. **153**(2–3), 303–317 (2005)

[Mu] K. Murasugi, On the braid index of alternating links. Trans. Am. Math. Soc. **326**(1), 237–260 (1991)

[Mu2] K. Murasugi, *On Closed 3-Braids*. Memoirs AMS, vol. 151 (American Mathematical Society, Providence, RI, 1974)

[Mu3] K. Murasugi, Jones polynomial and classical conjectures in knot theory. Topology **26**, 187–194 (1987)

[Na] T. Nakamura, Notes on the braid index of closed positive braids. Topology Appl. **135**(1–3), 13–31 (2004)

[Na2] T. Nakamura, Braidzel surfaces and the Alexander polynomial, *Proceedings of the Workshop "Intelligence of Low Dimensional Topology"*, Osaka City University (2004), pp. 25–34

[Ni] Y. Ni, Closed 3-braids are nearly fibred. J. Knot Theory Ramifications **18**(12), 1637–1649 (2009). math.GT/0510243

[Oh] Y. Ohyama, On the minimal crossing number and the braid index of links. Canad. J. Math. **45**(1), 117–131 (1993)

[Or] S. Orevkov, Quasipositivity problem for 3-braids. Turk. J. Math. **28**, 89–93 (2004). Also available at https://www.math.univ-toulouse.fr/~orevkov/

[PV] M. Polyak, O. Viro, Gauss diagram formulas for Vassiliev invariants. Int. Math. Res. Notes **11**, 445–454 (1994)

[PV2] M. Polyak, O. Viro, On the Casson knot invariant. J. Knot Theory Ramifications **10**(5), 711–738 (2001). Knots in Hellas '98, vol. 3 (Delphi)

[Ro] D. Rolfsen, *Knots and Links* (Publish or Perish, New York, 1976)

[Ru] L. Rudolph, Braided surfaces and Seifert ribbons for closed braids. Comment. Math. Helv. **58**, 1–37 (1983)

[Ru2] L. Rudolph, Quasipositivity as an obstruction to sliceness. Bull. Am. Math. Soc. (N.S.) **29**(1), 51–59 (1993)

[Sc] O. Schreier, Über die Gruppen $A^a B^b = 1$, Abh. Math. Sem. Univ. Hamburg **3**, 167–169 (1924)

[Sq] C. Squier, The Burau representation is unitary. Proc. Am. Math. Soc. **90**, 199–202 (1984)

[SSW] D. Silver, A. Stoimenow, S.G. Williams, Euclidean Mahler measure and twisted links. Algebr. Geom. Topol. **6**, 581–602 (2006). math.GT/0412513

[ST] C. Sundberg, M.B. Thistlethwaite, The rate of growth of the number of prime alternating links and tangles. Pac. J. Math. **182**(2), 329–358 (1998)

[St] A. Stoimenow, On polynomials and surfaces of variously positive links. J. Eur. Math. Soc. **7**(4), 477–509 (2005). math.GT/0202226

[St2] A. Stoimenow, The skein polynomial of closed 3-braids. J. Reine Angew. Math. **564**, 167–180 (2003)

[St3] A. Stoimenow, On the crossing number of positive knots and braids and braid index criteria of Jones and Morton-Williams-Franks. Trans. Am. Math. Soc. **354**(10), 3927–3954 (2002)

[St4] A. Stoimenow, Coefficients and non-triviality of the Jones polynomial, J. Reine Angew. Math. **657** (2011), 1–55; see also math.GT/0606255

[St5] A. Stoimenow, *Diagram Genus, Generators and Applications*. Monographs and Research Notes in Mathematics (T&F/CRC Press, Boca Raton, 2016). ISBN 9781498733809

[St6] A. Stoimenow, The braid index and the growth of Vassiliev invariants. J. Knot Theory Ram. **8**(6), 799–813 (1999)

[St7] A. Stoimenow, Positive knots, closed braids, and the Jones polynomial. Ann. Scuola Norm. Sup. Pisa Cl. Sci. **2**(2), 237–285 (2003). math.GT/9805078

[St8] A. Stoimenow, Knots of (canonical) genus two. Fund. Math. **200**(1), 1–67 (2008). math.GT/0303012

[SV] A. Stoimenow, A. Vdovina, Counting alternating knots by genus. Math. Ann. **333**, 1–27 (2005)

[SW] D. Silver, S. Williams, Coloring link diagrams with a continuous palette. Topology **39**, 1225–1237 (2000)

[Th] M. B. Thistlethwaite, On the Kauffman polynomial of an adequate link. Invent. Math. **93**(2), 285–296 (1988)

[Th2] M.B. Thistlethwaite, A spanning tree expansion for the Jones polynomial. Topology **26**, 297–309 (1987)

[Tr] P. Traczyk, 3-braids with proportional Jones polynomials. Kobe J. Math. **15**(2), 187–190 (1998)

[Vo] P. Vogel, Representation of links by braids: a new algorithm. Comment. Math. Helv. **65**, 104–113 (1990)

[Wi] R.F. Williams, Lorenz knots are prime. Ergodic Theory Dyn. Syst. **4**(1), 147–163 (1984)

[Wo] S. Wolfram, *Mathematica — A System for Doing Mathematics by Computer* (Addison-Wesley, Reading, 1989)

[Xu] P. Xu, The genus of closed 3-braids. J. Knot Theory Ramifications **1**(3), 303–326 (1992)

[Y] S. Yamada, The minimal number of Seifert circles equals the braid index. Invent. Math. **88**, 347–356 (1987)

[Yo] Y. Yokota, Polynomial invariants of positive links. Topology **31**(4), 805–811 (1992)

Notation

pages of introduction of some common notation

Symbols

$*$, 10

$A(D)$, 8

B_n, 9

F, 7

$G(A)$, 8

$IG(A)$, 9

$M(p)$, 84

$MFW(L)$, 12

$P(v, z)$, 6

Q, 8

$V(t)$, 7

$V_i, \bar{V_i}$, 8

W_Y, 65

$X(q, \lambda)$, 65

$[123\ldots]$, 10

$[23]+$, 10

$[X]_d$, 6

$[\beta]$, 10

$\chi(G)$, 9

$\chi(L)$, 6

χ_s, 6

$!$, 10

$!D, !K$, 6

∂H, 95

π_Y, 65

ψ_n, 30

σ_i, 9

$\sigma_{i,j}$, 11

Δ, 64

$\Delta(t)$, 7

$\Lambda(D)$, 7

$@$, 10

$\wedge$, 73

$b(L)$, 11

$c(D)$, 7

$c_\pm(D)$, 7

$c_\pm(\beta)$, 10

$g(D)$, 6

$g(K)$, 6

g_s, 6

l_{ij}, 31

max deg, 6

min deg, 6

$n(D)$, 6

$\vdash$, 66

$||p||_n$, 84

$s(D)$, 6

$\Delta(D)$, 9

$w(D)$, 7

© The Author(s) 2017
A. Stoimenow, *Properties of Closed 3-Braids and Braid Representations of Links*,
SpringerBriefs in Mathematics, https://doi.org/10.1007/978-3-319-68149-8

Index

The boldfaced entries refer to the page of definition (or most detailed introduction, or most basic reference). Some notions are used very frequently, and only the more important occurrences are recorded. Mentions are also often avoided if the term occurs in the title of the (sub)section.

A

algorithm
  braid conjugacy, 1
  Orevkov's, 4
  Schreier's, 1, 21
  Seifert's, 6
  Vogel's, 12
  Xu's, **12**, 15, 16, see also Xu's normal form
  Yamada's, 12

B

bicolar, **95**
Birman dual, **24**, 64
bound
  Morton-Franks-Williams (MFW), **12**, 16, 25, 39, 40, 42–44, 48, 57–60, 62, 75, 76, 79, 83, 90
braid, v, **9**
  A/B-adequate, see braid,semiadequate
  adequate, **11**
  almost positive, **10**, 11, 54
  alternating, 18, 21, 59, 60
  band representation, **11**, 94
    k-almost strongly quasi-positive, **12**
    almost strongly quasi-negative, 36
    almost strongly quasi-positive, **12**
    band-negative, see braid, band representation, strongly quasi-negative

band-positive, see braid, band representation, positive
  minimal genus, 2, 11, 16
  negative, **12**
  positive, 3, **12**, 15, 16, 25, 29, 52, 54
  strongly quasi-negative, 28
  strongly quasi-positive, see braid, band representation, positive
  strongly quasi-signed, **12**, 30
cabled, **50**
exponent sum, **10**, 16, 20, 23–26, 28, 30, 33, 40, 63–67, 69, 70, 72, 76, 77, 79, 80, 83, 88, 89, 91
index, **2**, 3, 4, **11**, 12, 20, 25, 26, 60, 61, 83, 90
  of alternating links, 4, 57
  of positive braid links, 4, 39
negative, **10**, 18
positive, v, **11**, 39, 54
pure, 31, 82
scheme, **42**
Schreier vector, **10**, 32, 33
semiadequate, **11**, 32, 33, 35–37, 40, 49, see also diagram, semiadequate
strand numbering, **31**, 94
word
  (syllable) extension, **10**, 19, 42, 46, 48–50, 54
  (syllable) reduction, **10**, 23, 24
  complexity, **44**, 46

© The Author(s) 2017
A. Stoimenow, *Properties of Closed 3-Braids and Braid Representations of Links*, SpringerBriefs in Mathematics, https://doi.org/10.1007/978-3-319-68149-8

Printed in the United States
By Bookmasters